Using data

a handbook for students of Business, Economics and the Social Sciences

with detailed guidance on the design of an investigative project

Jean Mangan

The Business School
Staffordshire University

First published 1995

ISBN 1 872849 40 7

Printed by York Publishing Services

Acknowledgement

The author is grateful to the following for permission to include data in this book. Crown copyright data from the "Monthly Digest of Statistics", "Economic Trends", "Regional Trends" and "Social Trends" are reproduced by permission of the controller of Her Majesty's Stationary Office (HMSO). No responsibility for the accuracy or comprehensiveness of the data is accepted by the Controller of HMSO.

Data from *British Historical Statistics* by B.R.Mitchell are reproduced by permission of Cambridge University Press. We are also grateful to the World Bank for permission to reproduce data from its "World Tables" and "World Development Report" and to the Statistical Office for the European Community, EUROSTAT, for permission to include data from its Chronos database.

CONTENTS

Data supplement

Preface

Statistics are used all the time – in newspapers, on television, in books, by politicians and by teachers in educational establishments. They are used to elaborate and support various arguments or views of the world. They are sometimes referred to as '*the facts*' and treated with deference. At the other extreme you may have heard the phrase '*lies, damn lies and statistics*'. Statistics have to be treated with caution – but they are very useful. They may be necessary in developing a case, but they have to be used appropriately. Statistics can be misused, or more commonly misinterpreted or poorly applied. The aim of this book is to provide the tools which will enable you to interpret and use descriptive statistical data meaningfully.

Data is used in many circumstances. There may be a reference in a sentence, or a fuller set of data presented in tabular or graphical form. You may come across summary statistics such as means and correlations or you may be set data interpretation exercises or required to use data in a project.

The first two chapters of the book are an introduction to the use of data. Chapter I covers the use of data in projects and reports. It provides a framework for planning a project and goes on to consider primary and secondary data collection and give some essential rules for presenting statistics. Chapter II covers basic data interpretation skills, dealing with the importance of units, types of series and, most importantly, with how to describe and analyse a data series.

Chapters III to V are concerned with the techniques which are used in analysing and presenting statistics. These start with the interpretation of graphs. Graphs are used very frequently in presenting statistics and it is essential to appreciate the different types of graph. This is followed by a chapter on how to interpret particular kinds of series such as percentages and indices. The last of these chapters covers descriptive summary statistics such as means and standard deviations.

The wide and important topic of causal relationships between variables is considered in Chapter VI. This deals with general issues such as the direction of causation between variables and using graphs at the start of an investigation into establishing relationships between data sets. It also deals with more specific aspects of trying to estimate relationships. The last chapter deals with questionnaire design. In projects where primary data has to be collected, the questionnaire design is of great importance. This chapter illustrates some good and bad practice.

Each chapter is subdivided into small manageable units which include worked examples. Questions are provided to give practice in the techniques, and also to encourage discussion. These questions are shown in *italics*, and the data is provided where appropriate.

More extensive data for use in investigations is provided in the data supplement. At the end of each chapter is a STUDY which gives a series of questions using tables from this supplement. The data provided is quite extensive to enable teachers, if they wish, to set up their own investigations. I am indebted to Stephen Hurd for the data and 'studies'. The data is also available on disk, with accompanying book *Basic Data Handling Techniques* by Stephen Hurd.

Jean Mangan
Business School
Staffordshire University
January 1995

I. USE OF DATA IN PROJECTS AND REPORTS

1. Planning a project – hypotheses and all that!

The need for an hypothesis

The basic use of data is to argue a case. It adds important detail, and is often viewed as evidence in favour of, or against, a particular proposition. The use of data can move an argument along from being a pure assertion to being a supported case.

Numbers can be given to the item under consideration, adding detail. For example, instead of the general statement that the UK housing market slumped in the late 1980s, an examination of the data allows more specific details.

'The number of property transactions in England, Wales and Northern Ireland fell from 223,000 in August 1988 to 115,000 in August 1991.'

The use of figures allows a quantitative assessment of the difference between two items, by comparing levels or rates of growth over either time or space. Statistics are useful to describe certain states, such as the annual rainfall in different regions. This information may be of interest in itself, or may help in identifying trends, relationships and problems.

Advancing an argument in the abstract is not sufficient in most circumstances. Support is needed for the case being presented, and this is often found in the use of data. Presentation of data can show a well-researched piece of work, rather than one just based on supposition. An example is the argument that the rise in interest rates in the late 1980s affected retail sales. The case can be made more convincing by the use of data:

'In the UK retail sales rose by 5.1% in 1987, when the interest rate (as measured by the Treasury Bill Yield) stood at 8.4%. In 1990, when interest rates had risen to 13.5%, there was zero growth in retail sales.'

Developing from these general uses of statistics, most pieces of research work are based on the notion of an hypothesis. An hypothesis is a falsifiable statement that may be arrived at through some complicated theory, or may be a simple statement of a particular relationship; for instance 'as incomes rise, sales of golf clubs rise'. In considering the uses of data in relation to hypotheses there are two distinct approaches:

Either

- collect all the data possible and then form an hypothesis

 or

- form an hypothesis and then use the data to test it.

Neither of these approaches in their pure form is tenable. Collecting all the data in a general area of study is too vast a project. Even if achievable, the amount of data collected would be too overwhelming to enable you to form a sensible hypothesis without a great deal of sorting of the material – and sorting itself implies some elementary hypothesis formation. Forming an hypothesis without any prior knowledge is equally untenable, some information on 'the world' is required to consider what relationships it may be sensible to develop.

During a course of study ideas, arguments and probably formal hypotheses will have been covered. Thus the relationships and hypotheses to be considered in any report or project will probably be suggested by the course material which you have studied. You may need to extend these during your investigations, using your accumulated knowledge of the subject.

In using data to test hypotheses, even very formally, the result is often inconclusive. If you find some data that does not fit the hypothesis then it is refuted. It is a false hypothesis. However, data cannot prove an hypothesis since there may always be some data, which you have not yet collected, which disproves the hypothesis. Data may support, but not prove, an hypothesis.

The data part of a project should never be viewed separately from the rest of the project. The task should be approached in an integrated fashion. Do not collect all the data you can easily find on the general area you have specified, and then include all of that data in the report in order not to waste it! This leads to muddle and poor presentation of a project. The correct approach is to be selective, decide at each stage what is important and what is not, for your particular investigation. The section below lists various factors that should be taken into account.

Aims of the project

Before starting any data collection you need to consider as clearly as possible the aims of the project and the approach you intend to take in pursuing these aims. There are various aspects to bear in mind:

- It is not sufficient to name a general area for investigation. The specific aims must be clearly laid out to give a feasible size of project. It is a common fault to be too ambitious in the breadth of a study and fall down on depth.

- Usually a project is required to develop and illustrate a part of the course of study. Techniques of analysis that have been covered during the course should be used to develop a logical argument. Rarely are assessors looking purely for description of an area, usually an analysis is required of the situation. What aspects of the course can be called on, particularly in terms of a framework and methods of analysis, should be considered at the initial stage.
- Many successful projects start by outlining an hypothesis for investigation e.g. higher prices will reduce sales.
- Statistics need to be used with care. This book describes how statistics can be used in a descriptive manner to support an hypothesis. That is whether, in general terms, the data seem to support the initial hypothesis. Testing hypotheses using formal statistical techniques is not covered in this text.

By defining the purpose clearly, the data you need should be narrowed down, allowing you to approach the data collection in an ordered manner.

Resources available

There are some general decisions to be made at the beginning of a project. What resources are available to you? In particular you need to consider the sources of information, processing facilities, and, most particularly, the time available to you.

Do you intend to use published materials? Most projects at least start with these. You may need to start with a library search to obtain suitable references. There are special catalogues available for such investigations. Are the important sources readily available in your library? You may need to order specific references, and time must be allowed for this.

Is information you require available through the media? Do you need to examine particular newspapers, watch certain television or radio programmes? Do you intend to write to particular organisations for information, or conduct your own surveys? This will all need careful planning.

What resources are available to help you analyse and present the data you collect? Have you a computer you can use, and will this help? If so, what software is available, and can you use it? If you are working on a group project you need to consider the resources of the group. Who is best suited to which task? But do make sure that everyone shares their specialist knowledge with others in the group.

What time scale are you working under? The project or report needs to be planned in terms of the time available, taking account of your other work. An aspect that often brings a problem is underestimating the time needed to write up a report. Be wary of this and allow yourself extra time.

Further stages in an investigation

Once you have a well-designed project outline, the initial data collection and examination can begin. However, it is rare for the project not to need modification as the investigation progresses. It may become clear that the topic is too vast, or that the data required is not easily available, or indeed not available at all. There are large amounts of data available today, but often the specific data that you need is missing. The aims and objectives may need modifying to take account of what data is actually available.

Set out below, briefly for the purpose of illustration, are the first stages of a project outline. In practice this should be developed in considerably more detail, but the example does illustrate some of the salient points.

general area	an industry study
chosen area	the British coal industry
specific aim	to investigate the hypothesis that there was increased competition in the coal market in the late 1980s and early 1990s
tools of analysis	demand and supply
information sources	library materials, newspapers, contact British Coal

In this example, the project could have been chosen from a very wide area of possible industry studies. Choosing a specific industry narrows this, but certainly not sufficiently for a viable project. The narrowing down of the area to recent market changes is a step in the right direction. It is now clearer what type of information will be of use, and what will be irrelevant. The indication of the use of demand and supply analysis does two things. Firstly, it further defines the scope of the project, as previously the study could have had a marketing rather than economic slant. It also enables the study to be analytical rather than purely descriptive. Finally, the initial information sources have been identified.

The project outline here is not sufficient, but it provides a starting framework for the investigation. It then needs developing with an outline of the main sections to be covered.

2. Primary and secondary data collection

Secondary data collection

Despite the name, secondary data sources are usually investigated before primary ones. Secondary refers to that fact that the data has been collected and presented by someone else. If such data is readily available, it is usually cheaper and quicker to use the secondary source. In many cases, particularly when national data is involved (for instance on output and prices), it is simply inconceivable not to use the secondary sources. In the UK these are published, in the main, by the Central Statistical Office (CSO).

Collection of secondary data is part of what is generally referred to as *desk research*. A good way to start any project is to explore recent published work in the area. This may start with references that have been given during your taught courses and continue with a library search for materials.

For UK national data, important data sources are:

Annual Abstract of Statistics
Economic Trends (especially the Annual Supplement)
Monthly Digest of Statistics
Social Trends
Regional Trends
Employment Gazette
Financial Statistics

All of these are published by the CSO and available in good libraries. Also of importance here are the minutes of Parliamentary committees, the reports of Royal Commissions and other government appointed bodies such as the Monopolies Commission.

For international data, important sources are:

Eurostat: *Eurostatistics*
International Monetary Fund: *International Financial Statistics*
World Bank: *World Development Report.*

For statistics and commentary on economic events, good sources are the quality press (such the *Financial Times* and the main broadsheet newspapers). For more detailed analysis the *National Institute Economic Review* (published by the National Institute for Economic and Social Research) and *Economic Outlook* (published by the London Business

School) provide commentary three times a year. For more official views, but also containing a wealth of statistics on developments in the economy, there are the Treasury's *Annual Statement and Budget Report* (published annually) and the Bank of England's *Inflation Report* (published quarterly).

Primary data

For some projects primary data collection is essential. This is where you go out and collect information yourself. The most common method is to use a questionnaire, to gain peoples' responses to questions on matters of fact or opinion.

Questionnaires may be distributed by post, or used in a personal interview. The advantage of the former is the cheapness and lack of any interviewer bias. The disadvantages are that response rates are usually much lower than for personal interviews and the difficulty of making the questionnaire 'foolproof' to the respondent. Clarifying questions cannot be asked of, or by, the interviewer.

For many types of information, primary data collection is the only possibility. The information simply does not exist elsewhere. However, such investigations should not be carried out lightly. They require careful planning and are often very time consuming to carry out properly. Always be careful to establish the aim of the survey. Identify the specific information that needs to be found out and allocate a proper time scale to the survey. Investigations done on the basis that they might help, or to pad out the report, are very rarely worth undertaking at all.

Usually a survey is piloted on a small scale, to check both that the information being collected is that which was intended and to help distinguish what specific information is of use and can be reasonably collected. The important aspect of questionnaire design is covered in chapter VIII.

3. Data use and presentation

When the information has been collected, the material requires analysing and presenting. The tools of analysis will usually have been taught in the area under study; how you intend to use these tools should have been considered in the initial stage. Specific ways of using and presenting data are covered in the later sections of this book. This section covers basic issues of concern whenever data is used, no matter how simple or complex, whatever the subject or purpose.

Categorising material

Material needs to be categorised into essential, supplementary and irrelevant.

- **irrelevant data**

Although careful planning at the design stage of a project will minimise the collection of irrelevant material, often unimportant data will have been collected, perhaps as part of a larger table with some relevant columns. Surplus material should be discarded. Do not include large tables where only a small section is relevant – extract the relevant material and discard the rest.

- **supplementary material**

This adds useful information but is subsidiary to the main argument. It should be placed in appendices. Placing this in the main text will detract from the main argument. Any material placed in appendices should be referred to at the appropriate point in the text, so that readers can turn to the appendices if they decide the additional information would be useful. Use appendices sparingly – always consider whether the material really falls into the irrelevant category. Appendices are also useful when material will be referred to several times at different stages of the report.

- **essential material**

Some statistical material will be essential for the case being argued. This should be included in the main body of the report at the appropriate point. Its importance should be clearly highlighted for the reader.

The form of statistical presentation

In writing up a report, remember that most of us find statistics rather indigestible. The reader must not be switched off, since the statistics are necessary to support the argument and present a strong case. Thus great care must be taken to include the most essential statistics and present these in an easily interpretable form. For many statistics this means using various graphical and pictorial modes of presentation. Summary statistics, such as the means of different groups, or moving averages, may also prove useful. These techniques are dealt with in later sections of this book.

The labelling of statistics

All statistics, whether presented in graphical or tabular form, should be clearly and fully labelled. There should be:

- a heading
- column and row headings
- clear labelling if more than one series is graphed or included in the table
- footnotes giving details of headings, changes in series etc.
- details of sources

Integrating the statistics

It is necessary, in the text of a report, to refer readers to a table or graph and tell them why they are looking at it. You are arguing a case. Make sure the reader knows your conclusions; do not leave him or her to search for them. Consider carefully the layout of the report and place the graph or table as close as possible to the text which refers to it.

Summary of Key Points

1. Specify the hypotheses which are to be investigated.
2. Check what published data is already available.
3. Be ready to modify your aims to take account of the data availability.
4. Consider carefully whether you need to collect any primary data.
5. Be ruthless in selecting data for inclusion in your report.
6. Label data carefully and always give the precise source.

STUDY A

Suppose you have been asked to carry out a research project on 'Newspaper Readership'. Prepare a plan of how you would do it, using the ideas outlined in this section.

Tables A1 and A2 of the Data Supplement contain some secondary source material. How might you use this in preparing your project?

II. BASIC DATA INTERPRETATION SKILLS

Describing a data series is necessary both in developing and in writing up your own report or project. This skill may also be useful on your course in essay writing and data interpretation exercises. The aspects explored in this section are general points that relate to data series. Specific items of importance are considered later.

1. Making sense of numbers – units

A data series may be presented to you in the form of a table or a graph. In analysing a data series, the first things to examine are the headings of the data and the units in which the data are described. In Table 1 on income below, are the figures for national income or weekly average family income? Is it for the UK, GB, or England? Are the units £s, £million, £billion or what? Read carefully the labelling of the statistics. The figures in the Table 1 are very different, but all are estimates of UK income in 1993.

Table 1: UK Income, 1993

Year	1	2	3
1993	572,027	458,361	103.5

Source: Monthly Digest of Statistics

In this case the data in each column is:
Column 1. Personal Income before Tax, £ millions
Column 2. Personal Disposable Income, £ millions
Column 3. Index of Real Personal Disposable Income (1990=100)

'Disposable' means after tax. Real means after adjusting for inflation. Both this and index numbers are explained further in Chapter IV. The important point here is that care must be taken in identifying data and establishing the most relevant statistics.

Sometimes it will be necessary to read not just the title and the units, but also the smaller notes at the bottom of a table to understand what is being shown. An example of this is the series given in Table 2. The heading and units are as given by the Central Statistical Office. An interpretation of the table could be that 38% of all 19–20 year old females were in higher education in 1989/90. However this is not what the figure shows. The table gives the percentage of full-time students by age. That is 38% of female students in higher education are aged 19–20. Always consider carefully what is being shown, whether there is more than one possible interpretation and whether the one you have chosen seems reasonable in view of the figures. Checking all the information supplied with the table should help in this.

Table 2: Higher Education – full time students by age, 1989/90, UK

Females	%
18 & under	17
19–20	38
21–24	27
25 and over	19

Source: CSO, Social Trends

2. Types of series

Next, it is important to establish what type of series you are dealing with. Below are some of the types you may encounter. The divisions are not mutually exclusive – a series may fit into more than one category.

(a) Cross section

This type of data series gives details of different categories of a particular item at a point of time. An example of a cross sectional series is the percentage of full time students by age given in Table 2 above.

(b) Time series

This type deals with the movement in a variable over time. Table 3 gives an example of time series data – quarterly data on new car registrations in GB for 1991 and 1992.

The table gives both the raw and seasonally adjusted data, and both are commonly available for quarterly and monthly national data. Which series should be used depends on the purpose of the investigation. If identifying the seasonal pattern is important, then the 'not seasonally adjusted' figures need to be used. If you are concerned with trends over a number of years the 'seasonally adjusted' figures may be more appropriate.

(c) Definitional

This type gives the breakdown of some aggregate variable into its components. Table 4 shows the breakdown of the UK current account of the balance of payments for 1993. It is divided into visible trade, which is the trade in goods, and invisibles, which is largely the trade in services.

Table 3: New Car Registration, GB

Year		000s	
		Not seasonally adjusted	Seasonally adjusted
1991	Q1	611	420
	Q2	452	373
	Q3	712	419
	Q4	359	385
1992	Q1	539	382
	Q2	481	387
	Q3	709	405
	Q4	404	425

Source: Economic Trends

Table 4: UK Balance of Payments, Current Account, £mn, 1993

Current account	£mn
exports	120,839
imports	134,519
visible balance	-13680
invisible balance	2753
current balance	-10670

Source: Economic Trends

One point to note with this type of data series is that the categories should approximately add up to the aggregate. Here, exports minus imports gives the visible balance, and the visible balance plus the invisible balance gives the current balance. It is only approximate because in some cases rounding of the figures will leave a minor difference.

(d) Multiple series showing related variables

Often two or more sets of data need to be presented together to illustrate a point. The data may be presented in the form of a table, a time series plot of the two variables on one graph, or in the form of a scattergram (also known as an XY plot). In this case always look carefully to see if the scale is the same for both variables. An example of two related variables is income and consumption; the relationship between these, after adjusting for inflation, is shown in Graph 1. In this case the scale on both axes is the same for both items.

Graph 1: Real Personal Disposable Income and Real Consumption, 1980–93 £bn 1990 prices

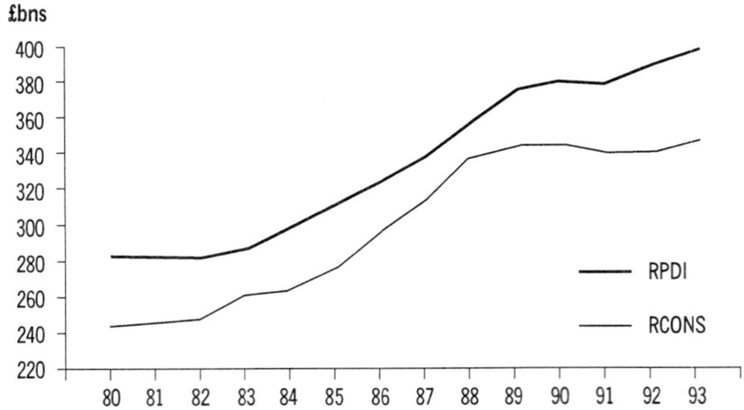

Source: Economic Trends

Exercise 2

Identify into which of the above types (a) to (d) the data in (i) to (iv) below can be categorised. Note that a table may fall into more than one category, they are not mutually exclusive.

(i) Population in millions,

Year	Total	Male	Female
1951	44	21	23
1961	46	23	24
1971	49	24	25
1981	49	24	25

Source: B.Mitchell, British Historical Statistics

(ii) Population of Towns, England and Wales, 000s

Year	Southend	Bath	Wigan
1801	–	33	11
1851	–	54	32
1901	29	50	82
1951	152	79	85

Source: B.Mitchell, British Historical Statistics

(iii) Working Population, 1989 millions

Year	Millions
Japan	63
USA	124
EC	143

Source: Eurostat

(iv) Retail Sales and GDP, UK, 1990=100

Year	Retail Sales	GDP
1988	97.3	97.3
1989	99.3	99.4
1990	100.0	100.0
1991	98.9	97.7
1992	99.5	97.2

Source: Economic Trends

3. Describing and analysing a data series

Time series data

The data may be for years, quarters, months or some other period of time. In describing time series data it is not sufficient to make vague general statements, such as 'there is an upward trend' or 'the value has increased over the last 10 years'.

Details to consider include:

- **trend**
  In describing such a data series the first thing to identify is the overall trend. Over the total period considered has the value of the item gone up, down or remained stable?

- **turning points**
  Can different periods be identified i.e. are there some periods when the item is increasing in value and periods when it is decreasing? When does the data change direction? How many times does this happen?

- **magnitude**
  Can any indication be given of the magnitudes? For time series this will be the changes over time. These details may be described in actual numbers, or in words that correctly project the relative sizes, e.g. percentage changes.

- **seasonality**
  Is the series seasonally adjusted? Many economic series vary with the season e.g. for many goods the highest sales are in the run up to Christmas. If the data has been 'seasonally adjusted' an attempt has been made to iron out these seasonal fluctuations so that the longer term trend can be more easily distinguished.

- **hypothesis**
  In describing a series, draw attention to aspects which throw light on the area under investigation, or to a specific hypothesis, if possible. This is important both when interpreting data which you have been given, and in using data in a project which you are developing. You need, when describing data, to be aware of the relevant analytical framework from the course you are undertaking. This helps to identify the important aspects to look for in the data.

This example uses the data illustrated in Graph 1. Using the points made above, we can say that both consumption and income increase over the period. However, in the the period 1986–89, consumption was growing faster than income (the gap between the two lines on the graph is narrowing). In the years 1990–93 income grew faster than consumption (the gap between the two lines on the graph is widening). Consumption and income grew by approximately £100 bn. over the period 1980–1993. The growth was particularly great in the period 1986–88.

The initial hypothesis to consider is the consumption function. In this hypothesis, consumption increases as income increases. This fits with the data in Graph 1. However, the simple form of the hypothesis, as commonly found in introductory textbooks of economics, has the relationship between the two as being linear. This cannot be true here, as consumers spent more of their income in the period 1986 to 1988. This is shown by the lines on the graph becoming closer and needs to be further commented on. The period 1986–88 saw a consumer expenditure boom that may have been caused by the low interest rates of the period, and/or the liberalisation of the financial sector.

Exercise 3

Use the data on new car registrations in Table 2 to:

(i) *Describe the difference in sales between seasons and between the two years (Hint: for differences between seasons use the not seasonally adjusted data, for differences between the years use the seasonally adjusted series).*

(ii) *Give some indication of the magnitude of the changes you have identified in (i)*

(iii) *Can you think of any reasons to explain the patterns you have described in the car market data ?*

Cross sectional data

- **subgroups**
  Can different sub groupings be identified? The idea here is to split the data in order to consider several items together which appear to be related.

- **magnitude**
  Can any indication be given of the magnitudes? For cross sectional data the relative size of the various categories needs to be considered. These details may be described in actual numbers, or in words that correctly project the relative sizes.

- **hypothesis**
  As with time series, reference to an hypothesis in examining the data is important. Again this will generally be drawn from the course you are taking.

Table 5 gives an example of cross sectional data. The figures are for the rate of redundancies per 1000 employees in 1993. The table shows that the highest rate was in the North, followed by the West Midlands, with the South East having the lowest rate. At a more detailed level it shows that the rate was nearly one third higher in the North than in the South East. This difference may be related to the state of the economy in the three regions, and/or the different types of industry in the regions.

Table 5: Redundancies, rate per 1000 employees, 1993

Regions	Rate per 1000 employees
North	16.5
West Midlands	13.9
South East	11.3

Source: Regional Trends

Summary of key points

1. The most important first step in interpreting data is to examine carefully the headings and units of the data.
2. Cross sectional data gives details of items at one point in time and is useful for making comparisons between categories. In describing such data, consider the magnitudes of the different items and possible sub-groupings of the data.
3. Time series data gives changes over time. The trend, turning points and magnitudes at different points of time are important in describing this kind of data set.
4. A major use of data is to examine relationships, so many presentations of data show more than one item.
5. In analysing data, either cross sectional or time series, reference needs to be made to an hypothesis. The important point to establish is whether the data supports or enables you to refute the hypothesis.

STUDY B

Say as much as you can about the data in Tables B1, B2, B3 and B4 of the Data Supplement. Include comments on:

(a) the type of data
(b) the units of measurement
(c) trends, turning points, magnitude and seasonality for time series, subgroups and magnitude for cross sections.

III ANALYSING AND PRESENTING DATA USING GRAPHS

Introduction

Graphs illustrate orders of magnitude, relative sizes, trends and relationships among variables. They are useful in that it is easier to assimilate the information in this form than through a table of figures. The eye can more easily appreciate the situation from a good diagram. For this reason information will often be given to you in graphical form and if you are using data to argue a case graphical presentation is a useful tool.

In using graphs several basic things need to be borne in mind:

- **heading**

  The headings of the graph should describe clearly and concisely what the graph is showing.

- **the axes**

  The axes further define what is being illustrated and the labelling should include the units of measurement. For instance, if the graph is to show temperature variation in a particular region over the year, the X axis may give the months of the year and the Y axis the temperature in °C. In reading or constructing a graph the sequence is from the X (horizontal) to the Y (vertical) axis in interpreting the data.

- **data source**

  The source of the data should always be given, to give authority to the figures and to enable them to be checked.

- **interpretation**

  In interpreting a graph, the relevant details of the data need to be carefully considered. As with any examination of data this is best done by reference to an underlying framework or hypothesis. With a graph showing changes over time, the trends need to be carefully considered. The interpretation may need to cover both qualitative changes (that is ups and downs) and also quantitative changes (that is the numbers involved). With comparisons using cross sectional data, it may be necessary not just to identify the smallest and the largest items, but again some consideration of the size of the differences may be important.

- **graphical form**

Before constructing a graph consider what type of graphical form best suits the material. The following descriptions of the types of graphs give some ideas and some do's and don'ts. In general terms the illustration should be chosen to illustrate the point being made in the simplest manner. If several graphs are to be used, a variety of forms may help enliven the presentation – but only if the alternatives are appropriate.

- **size**

The size of the graph should be sufficiently large for the reader to appreciate clearly the salient details. It should be sufficiently small to fit onto the paper at the appropriate point for reference. The graph, the description and use of the material in arguing a case should be close together in the text. Referring back and forth for information is difficult and annoying to the reader.

There are several main types of graph. In the rest of this section the major forms are dealt with briefly and examples are given of the points listed above. These graphs can be drawn fairly easily using many graphics options in computer word processing and spreadsheet packages.

1. Line graphs

Description

Line graphs plot the relationship between two or more series by way of connected data points. They are suitable where the material is continuous i.e. where it makes sense to interpret the points in between the data points (technically to 'interpolate'). Line graphs are commonly used for time series – where we have a data series that runs over a number of time periods, for instance, months or years.

Construction

In drawing a line graph, the first decision is on an appropriate scale for the axes. This depends on the units of measurement and the size of graph that is considered appropriate. The axes should be carefully labelled with the units of measurement: in Graph 2, the Y axis is in millions of tons and the X axis is in years. The points are then plotted onto the XY space and the data points are connected to give the line graph. For instance, the output for 1905 was 236, and this data point is plotted 236 units vertically above 1905. Always be careful to give a heading that describes what is being measured, in this case the annual output of coal 1900–1945.

Graph 2: Coal output, Great Britain, 1900-1945

Source: B. Mitchell, British Historical Statistics

Interpretation

In interpreting such a graph first consider the title and then the scales of the axis. In interpreting the graph a point on the X axis is related through to the Y axis. Here the X axis is in years, starting in 1900, while the Y axis is in millions of tons of coal. The Y axis does not start at 0, but at 170 million tons. It is quite common with graphs to adjust the scale in this way, in order to highlight the fluctuations, but in drawing graphs take care not to exaggerate small fluctuations by inappropriately reducing the scale. In this case, the change over the period is more than 90 million tons and the adjustment of the axis is quite acceptable. In viewing the graph, do not make the error, however, of considering that coal output had fallen to nearly zero by 1945. Always refer closely to the axis. With this graph the following types of interpretation can be made:

- levels, e.g. coal production in the U.K. in 1900 was 225 million tons.
- peaks and troughs, e.g. the peak year for coal production was 1910 (the interpolation is not quite correct – it was in fact 1913).
- trend, e.g. coal production fell by over 30% in the period 1910–1945.
- trends in sub periods, e.g. coal production rose rapidly up to 1910, suffered a decline in the 1910–1920 period, and a more sustained decline post 1930.

If you are asked to describe a graph, remember to do so reasonably fully, trying to bring out the important aspects of the series. These can be identified by considering the data series together with any appropriate information or hypothesis that has been covered on your course. Look for general trends. Is the rate of increase or decrease itself changing? Are there clear turning points – peaks or troughs?

Multiple series

More than one series can be put on a graph. Usually, but not always, these will have the same scales. This is usually done for purposes of comparison between the series, to consider differences and similarities and to draw inferences about connections between the variables. This is part of the process of developing and investigating hypotheses. Graph 3 illustrates two series that are correlated, the percentage change of GDP and imports in the UK. GDP percent change is a measure of the overall growth in the economy. With such graphs each series should be clearly marked. A common way of doing this is to use a legend; this is the key to the series, which are usually drawn slightly differently. In Graph 3, import growth is shown by the thin line and GDP growth is the broad line. This is identified by the legend, given underneath the graph.

Graph 3: Percentage change in GDP and Imports, UK

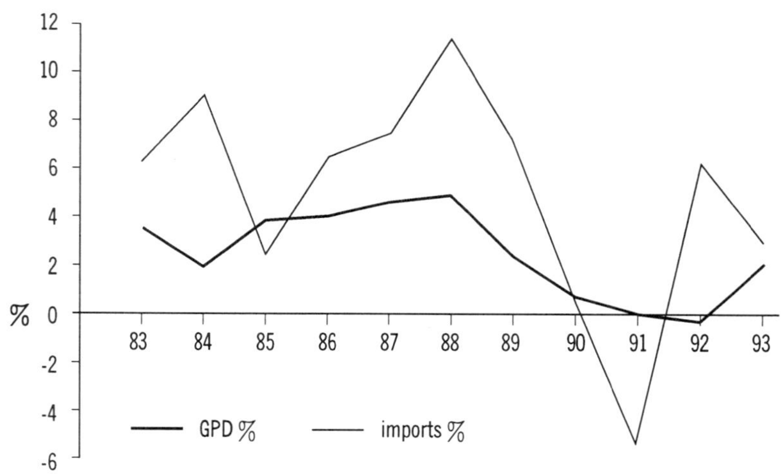

Source: Economic Trends

An introductory examination of this type of graph can cover the following types of comparison:

- **overall comparison**, e.g. import growth has generally been higher than growth in the economy
- **particular points in time**, e.g. when GDP growth has been particularly low, as in 1991, import growth has also been very low.

A more analytical approach would refer to an underlying hypothesis or theoretical framework that links these two variables. Here the hypothesis may be that import growth is positively related to GDP growth, but that other variables, such as the liberalisation of world trade, or exchanges rates, are also important in affecting import growth.

2. Bar charts

Description

A bar chart shows the data as a series of rectangles, the height of the rectangle giving the level of the variable. Bar charts are often used for cross sectional material, that is data at a point in time. The categories, for example, may be households, countries or industries. It is also an appropriate form to use for time series data when interpolation is dangerous because the gap in time is considered too long. An example of this could be a comparison decade by decade.

Graph 4: Regional unemployment rates, 1993

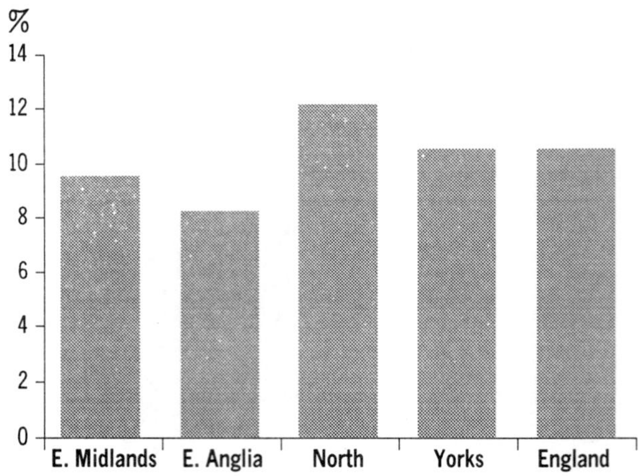

Source: Regional Trends

Construction

As with a line graph, the first consideration is the scale of the axis. On the X axis each category or period is given a certain length (this length will depend on the number of categories/periods, and the overall size required for the graph). The Y axis shows the variable of interest, in this case the unemployment rate, which determines the height of the bar. In Graph 4 the unemployment rate in several regions is compared to the English average for 1993. The first 'bar' is the East Midlands, and this is given a height of 9.5, this being the percentage unemployed in that region.

Interpretation

In interpreting a bar chart it is again standard to start with the X axis. The height of the bars gives the amount for each classification and this can be read off from the Y axis. It is easy to make quick comparisons by looking at the relative heights of the bars.

Interpretation includes:

- **level of a particular category**, e.g. the level of unemployment in East Anglia in 1993 was 8.1%.
- **comparison to the average**, e.g. of the regions covered, the North had the highest employment rate, and was the only one that was well above the English average.
- **comparison between categories**, e.g. unemployment in the North was higher than in East Anglia, or more specifically, unemployment was over 30% higher in the North than in East Anglia.

Complications

(i) Comparing years

A common use of bar charts involves placing different years side by side for each category. An example of this is given in Graph 5, which extends the previous example by adding the 1990 figures. It also gives the exact levels over each bar – a common method of adding detailed information. However, be careful not to add too much detail which may detract from the important message the graph is there to portray.

Graph 5: Regional unemployment rates, 1993 and 1990

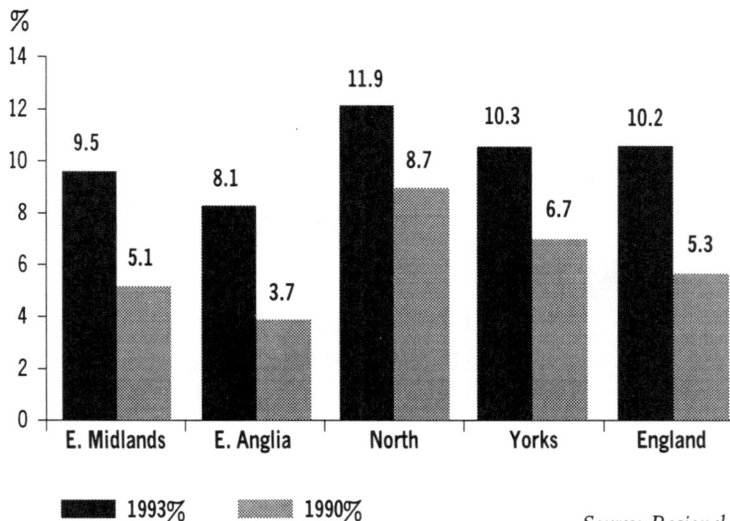

Source: Regional Trends

In addition to those considered earlier, the following kinds of comparison can then be made:

- **overall**, e.g. unemployment grew in all the regions considered over the period 1990–93.
- **specific category**, e.g. the increase was greatest in East Anglia, where the percentage unemployed more than doubled.
- **the ranking**, e.g. the ranking of the regions in terms of unemployment was the same in both periods.

(ii) Stacked bar charts

A stacked bar chart takes the various components of a variable and records these one on top of the other in a bar until the whole is reached. Graph 6 shows two stacked bar charts showing the stock of dwellings by tenure, in the U.K., in 1981 and 1992. The three categories add up to the total housing stock and the gain is more detail. Graphs of this kind can nicely illustrate:

- **the importance of different components**, e.g. owner occupation is the most important form of tenure in the United Kingdom
- **overall trends**, e.g. the stock of dwellings increased by over 2 million between 1981 and 1992
- **trends between categories**, e.g. owner occupation grew substantially over the period, while renting from Local Authorities fell.

Graph 6: Stock of dwellings by tenure

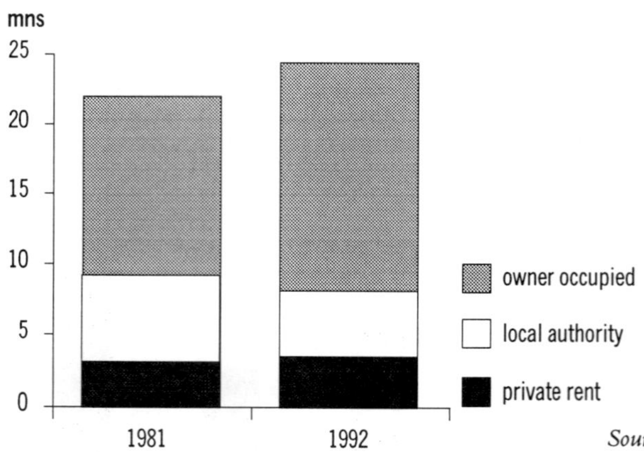

Source: Social Trends

However, this kind of graph cannot be used to illustrate minor differences between the categories. Also the order of the categories is important for easy interpretation. Here, putting the privately rented category, which has remained almost constant, on the bottom, enabled easier interpretation of the changes to the other categories.

A similar exercise can be done with line graphs, with the breakdown of an aggregate variable into its components. A variation is the proportional bar chart, as illustrated in Graph 7. In this the total bar is

100%, and the percentages of the components are shown. This version is useful in showing changes in the size of components when bar charts are compared in different years; it cannot, of course, show changes in the total amount of the variable. This type of graph would be the most appropriate if you wanted to illustrate how the types of tenure have changed in overall importance.

Graph 7: Percent of dwellings by tenure

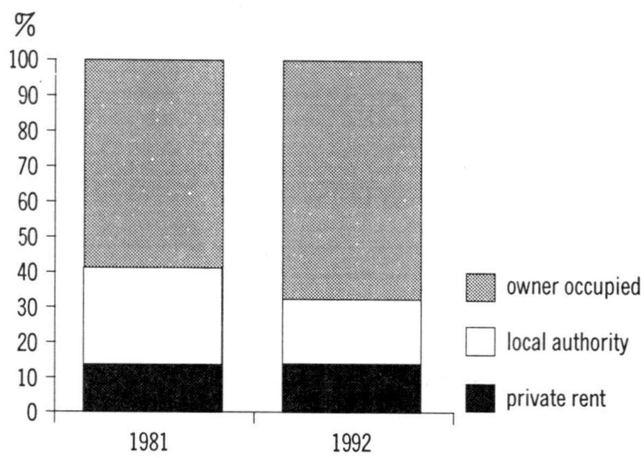

Source: Social Trends

3. Histograms

Graph 8: Frequency of interest rates, UK 1986 to 1993, monthly data

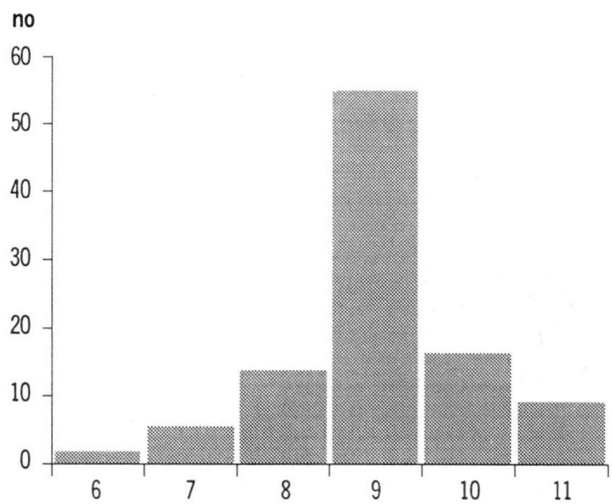

Source: Economic Trends

Description

A histogram looks rather like a bar chart. The difference is that with a bar chart the absolute amounts in each category are put on the Y axis, whereas a histogram is concerned with the frequency of events.

Construction

With a histogram the X axis is the class interval and the blocks are proportionate to the frequency. In Graph 8, the X axis is the interest rate and the Y axis is the frequency of the occurrence over the period 1986 to 1993 in the U.K. The data is monthly, so that there are 96 observations in total. Over the period there were 13 observations in the 8–8.99 range, and so the height of the bar is at 13; similarly there were 54 observations in the range 9–9.99.

Interpretation

A histogram is interpreted in terms of how likely an occurrence is. For example:

- **the most frequent level** e.g. the most frequent interest rate was in the range 9.00 to 9.99%
- **infrequent levels** e.g. the likelihood of a quarter drawn at random having an interest rate of less than 8.00% is low.

The interpretation of histograms can lead on to the consideration of a relevant theoretical probability distribution. However, that development is outside descriptive statistics and the coverage of this book.

4. Pie charts

Graph 9: Working population of the EU

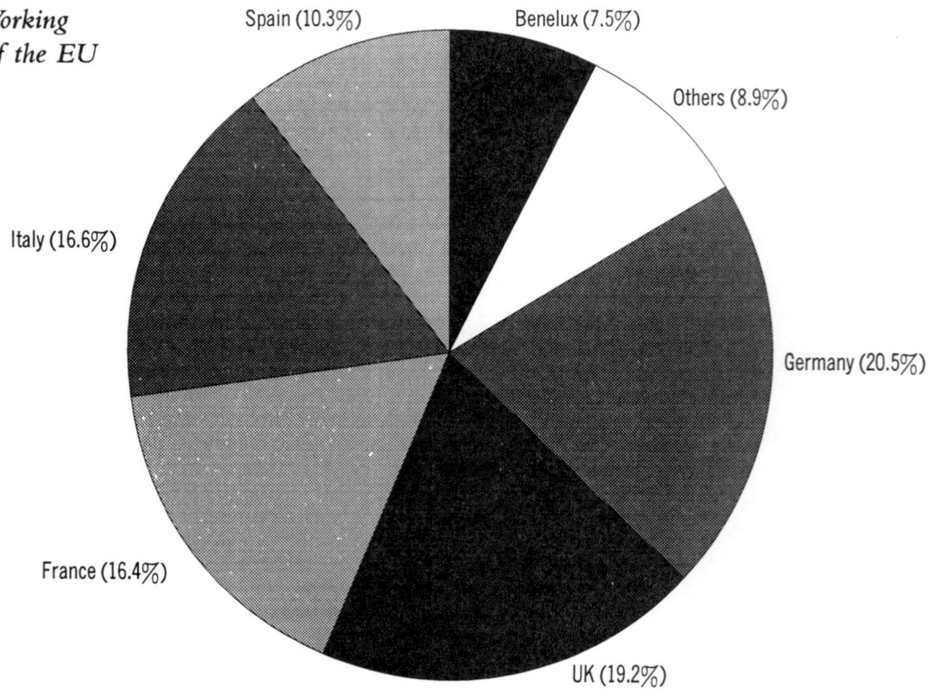

Source: Eurostat

Description

A pie chart is useful to show the division of an item among its components – the division of the pie into slices. For instance, the working population of the EU is split between the different countries in Graph 9.

Drawing

The total circle represents the whole item and the components are represented by the proportion in each category. Many computer graphics programs will draw pie charts automatically. To draw a pie chart by hand requires a compass and protractor. A circle has 360°, thus to divide the circle into the correct proportions requires the initial calculation of the proportions and multiplying these by 360. Having drawn the circle the degrees can then be marked off with a protractor. In the example shown in Graph 9, Germany accounts for 20.5% of the E.C. working population. The degrees in the circle thus allocated to Germany are:

$$360 \times \frac{20.5}{100} = 73.9^\circ$$

Interpretation

Interpretation focuses on the proportion or percentage in each category:

- **size of a slice**, e.g. the UK working population was close to 18% of the total.
- **comparison of slices**, e.g. the largest proportion of the EU working population was in Germany, followed by the UK.

Pie charts can also be used for comparison, by, for instance, considering a different year in the example above. The comparison then focuses on the differences in components shares of the total.

Exercise 4

The three tables (a), (b) and (c) below give data for which you should choose a suitable graphical mode of illustration. Comment on what is shown by the graph in each case.

(a) Working population, 1991

	Millions
EU	146
USA	125
Japan	65

Source: Eurostat

(b) Consumer expenditure on durable goods, 1990 prices

	Billions
1985	25.2
1986	27.9
1987	30.3
1988	35.0
1989	36.8
1990	34.7
1991	30.5
1992	30.5
1993	33.2

Source : Economic Trends

(c) UK employment (thousands)

Sector	1981	1991
Central Government	2419	2183
Local Authorities	2899	2969
Public Corporations	1867	747
Private Sector	17160	19826

Source: Economic Trends

SUMMARY OF KEY POINTS

1. Always label graphs fully and give the data source.
2. There is a wide variety of graphical forms available, including line graphs, bar charts, histograms, and pie charts. Be careful to choose a form which is suitable for the data to be displayed.
3. Use graphs to argue a case by fully interpreting what the graph illustrates in the context of the problem being discussed.

STUDY C

Given the data in Tables C1, C2, C3 and C4, choose a suitable graphical mode of illustration. Draw graphs for each table and comment on what is shown by the graph in each case.

IV ANALYSING AND PRESENTING DATA IN DIFFERENT FORMS

Data is often changed from the original, or raw, form in which it is collected in order to make it more understandable and to relate it to a particular purpose. This chapter covers the most common ways in which data is transformed. Some of the data you obtain or are given in a project may have been transformed in these ways. You may also want to transform data yourself in order to make it tie in more clearly with your argument.

1. The use of percentages

Introduction

A data series may be presented in a form that gives variables as a *level* or as a *percentage*. It is often much easier to interpret the percentage than the level, which may simply appear to the reader as a large number getting even larger or maybe smaller! If the data is included to illustrate a trend, or to show the differences between various groups, percentages are a better way of presenting the material to get the important point across to the reader.

Percentage in a category

■ **formula**

With cross sectional material percentages are worked out on the total to indicate the proportions in each category. The percentage in each category is found by taking the number in that category, dividing by the total and multiplying by 100 i.e.

$$\frac{\text{number in category}}{\text{total in all categories}} \times 100$$

■ **example**

Table 6 gives the kilometres of bad quality waterways in England and Wales and in the two regions. In the North West 265 kilometres are of bad quality, out of 614 kilometres in England and Wales, giving as a percentage:

$$\frac{265}{614} \times 100 = 43.2\%$$

Table 6: Water quality, England and Wales, 1988

Region	km of waterways				
	Good	Fair	Poor	Bad	Total
North West	3386	1334	915	265	5900
Thames	1597	704	117	0	2418
England & Wales	25984	9713	3465	614	39776

Source: Regional Trends

Interpretation

Percentages give an easy way of comparing the importance of each category. From the raw data it is possible to make simple deductions such as that the North West region has by far the longest length of bad quality water. Percentages are much easier and quicker to interpret and give preciseness. However, there is still a need to be careful in the interpretation. The North West had 43% of the bad quality waterways in England & Wales. However, it is **not true** that 43% of the waterways in the North West are grossly polluted. To obtain the percentage of North West waterways that come in the bad category, the percentage of the row is required, not the column, i.e. the 265 kilometres of bad quality water have to be compared with the 5900 total kilometres of rivers and canals in the North West, which gives 4.5%.

Exercise 4

Using Table 6 above:

(i) *calculate the percentage of waterways in the Thames area that are badly polluted*

(ii) *calculate the percentage of the badly polluted waterways in England and Wales which ares in the Thames area.*

(iii) *compare the North West and Thames regions using these statistics*

Rates of growth

formula

In the case of time series material the percentage change gives the growth in the variable over a specific time period, often a year. The rate of growth of a variable is calculated by taking the difference, dividing by the original and multiplying by 100, i.e.

$$\frac{(\text{new value} - \text{original value}) \times 100}{\text{original value}}$$

example

In Table 7, the first column gives the level of population in Greater London in the years 1984 to 1991. The second gives the rate of change, or growth of the population. For example, the calculation for the rate of growth of the population from 1984 to 1985 is:

$$\frac{(6767.5 - 6756.0)}{6756} \times 100 = \frac{11.5}{6756} \times 100 = 0.2\%$$

Table 7: Population of Greater London

Year	'000s	% Change
1984	6756.0	-
1985	6767.5	0.2
1986	6775.2	0.1
1987	6770.4	-0.1
1988	6735.4	*
1989	6756.4	0.3
1990	6794.4	*
1991	6803.1	0.1

Source: Regional Trends

Interpretation

If the concern is with the trends in a variable, with its growth over time, then the percentage directly illustrates this.

However, it is important to interpret such a series correctly . Between 1985 and 1986 the growth rate fell from 0.2 to 0.1%. This does not imply that the population of Greater London declined, but the rate of growth declined. The population was still increasing. A fall in the population is shown by a negative growth rate, as between 1986 and 1987 in the table.

Exercise 5

Calculate the percentage change between 1987/88 and between 1989/90 in Table 7.

What happened to population growth in the conurbation over the period 1985–91?

It is also very important when looking at percentage changes to consider whether the definition has changed over the period. Table 8 gives the population of Birmingham in the censuses of 1911 and 1921. The area of the city changed over this period. The two figures given for 1911 are:

(a) as shown in the census for the year concerned.
(b) as shown in that census for the area of the city in 1921

To compare like with like, in terms of area, it is the 1911 (b) figure that has to be compared with the 1921 (a) figure, which gives an increase of 9%. If, however the concern it with how the city has grown, including changes of the area, it the (a) to (a) change that is appropriate, which gives the much larger increase of 75%. The city boundaries extended over this period. Neither percentage change is correct or incorrect – but there is a definite need to be clear on what is being compared!

Table 8: Population of Birmingham, thousands

'000s			% change 1911/1921	
1911(a)	1911(b)	1921(a)	(a)/(a)	(b)/(a)
526	840	919	75	9

Source: B.Mitchell, British Historical Statistics

The figures above give the overall percentage change over the ten year period. It is also possible in examples such as the one above to calculate the annual compound rate of growth over the period. This is the rate of growth each year that would give the overall increase over the period. This is **not** equal to 7.5% (75/10) for the (a) to (a) change, because of the cumulative nature of growth. For instance, if the population of Birmingham grew by 5% between 1911 and 1912, growth in 1912 is measured from the base of 552.3 (i.e. 526[1+0.05]) not the original 526.

The annual compound growth rate is found by the formula:

$F = O(1+g)^n$ where F = final year level
O = original year level
n = number of years
g = growth rate (in proportion terms)

For the (a) to (a) change above this gives:

$$(1+g) = \sqrt[10]{\frac{919}{526}} = 1.0574 \quad g = 0.0574$$

which gives a compounded annual growth rate of 5.74% over the ten-year period.

2. Indices

Introduction

An index is a way of measuring the change in a variable from a given base point. This is useful when considering the change over a number of years, or when comparing in cross section to a chosen standard. The kind of question that can be answered is:

(i) how much did prices increase over the 1980s?

(ii) how does the standard of living in the UK compare to the rest of the EU countries?

Many published statistics are produced as indices as it is a common method of showing changes in aggregate statistics, that is where various categories have been added together to get the overall measure.

Calculation

The base is given the value of 100. Other points are calculated by the formula:

$$\frac{\text{original value of point of interest}}{\text{original value of base}} \times 100$$

In Table 9, the data in columns 1 and 2 is already in index form with 1980 as the base. In columns 3 and 4 the base year has been changed to 1985. The calculation for the new 1986 figure for the U.K. is done as follows:

$$\frac{300}{242} \times 100 = 124$$

Table 9: Index of share quotations

Year	1980=100		1985=100	
	UK	USA	UK	USA
1985	242	157	100	100
1986	300	199	124	126
1987	397	241	164	153
1988	358	222	*	*
1989	428	266	176	169

Source: Eurostat

Interpretation

An index gives the percentage increase or decrease from the base point. From the statistical viewpoint any year can be chosen as the base. The UK Central Statistical Office changes the basis for indices it calculates every 5 years. When calculating your own indices chose as the base the year you wish to compare from. The index then gives the percentage change from that point.

The example illustrates this point. The original data in columns one and two is already in index form, with 1980 as the base. This would be fine for considering changes since 1980. But what if you want to compare the relative performance of the UK and USA stock market since 1985? The comparison is difficult given the original form of the data. Once the base is changed to 1985 it becomes easy. Over the period, the UK stock market outperformed the US market, growing by 76%, compared to 69%.

Exercise 6

(i) *Fill in the two missing figures for 1988 in Table 9.*

(ii) *If, in 1980, you had invested money in the stock markets of the USA and the UK, which grew by the same rate as the stock index, by how much would the value of your investment have increased in the two countries by 1985?*

Weighted indices

An important concept is the weighted index. This provides an indication of the average changes in a group of related items. In the example above, the stock market indexes are weighted indices, giving the average change in certain share prices.

Perhaps the most commonly referred to weighted index is the Retail Price Index (RPI). The percentage change in the RPI gives the inflation rate. The calculation of the RPI involves three steps:

(i) a household survey to establish the relative importance of different goods in the average household budget. This produces the weights for the index.

(ii) a price survey covering the different categories of goods from different types of shops in different parts of the country.

(iii) the calculation of the index by taking the weighted average of the price changes from a given base point.

3. Real and nominal variables

Introduction

When studying trends in an economy over time, in variables such as GDP, exports and consumer expenditure, the variables are measured in money terms, basically because they are aggregates. Production, for example, is of cars, insurance services, oil etc. To obtain an overall measure we have to sum these very different kinds of item and for that money is the unit of account. The series is the money, or nominal, value of production. However, there is a major difficulty with using money as the unit of account as it has changed its value over time with inflation. To deal with this problem variables can be expressed in real terms, which weights the series to remove the problem of inflation.

Calculation

The method is to take a given base year and express the data in that years' prices. The calculation is:

$$\frac{\text{nominal value}}{\text{price index}} \times 100$$

and the result is the variable at the base year's prices.

Example

Table 10 illustrates this for the Gross Domestic Product (GDP) in the UK. The first column is the nominal data, the second is the price index, with a base of 1990. Real GDP is the first column divided by the second, multiplied by 100. For example, the calculation for 1991 is:

$$\frac{495}{105.8} \times 100 = 468$$

Table 10: Gross Domestic Product, U.K, at factor cost.

Year	Nominal GDP £bns	Price index 1990=100	Real GDP £bns 1990 prices
1990	479	100.0	479
1991	495	105.8	468
1992	514	110.5	*
1993	543	114.4	475

Source: Economic Trends

Interpretation

In official statistics the *real* GDP is referred to as GDP. '*at constant prices*' The *nominal* value is referred to as '*at current prices*'. The GDP for 1993 at 1990 prices is £475 bns.

Where variables are measured in money terms it is important to establish whether the nominal or real value is given by the data. Often the latter is of more interest because the concern is not with increases or decreases that are due to inflation, but with what has happened to the underlying trend, or real factors. Changes in the real value of a variable are an attempt to measure quantity changes. In the example above money GDP rose each year, but real GDP fell in 1991.

The Graphs 10 (a) and (b) further illustrate the importance of considering real rather than monetary data in many circumstances. Graph (a) is of real consumption and income, as examined in the section II.2. Graph (b) is of consumption and income in money terms. Most of the information on the income / consumption link is lost in (b)

Graph 10(a): Real Personal Disposable Income and Consumption 12990 prices

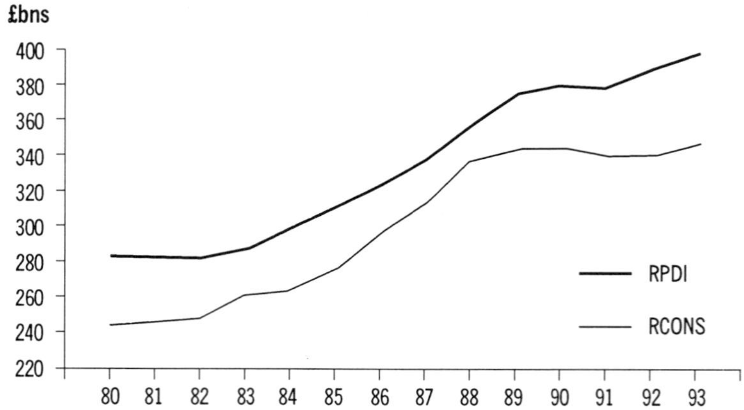

Graph 10(b): Money Personal Disposable Income and Consumption

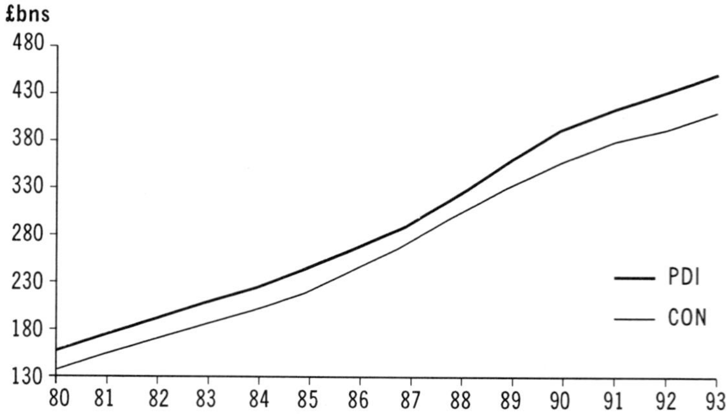

Source: Economic Trends

because of the overwhelming influence of inflation (which means that the scale of the Y axis has to be much smaller).

Exercise 7

Given Table 10 above:

(a) *calculate real GDP for 1992.*

(b) *identify years during which the real value of GDP fell. Did the nominal value fall as well in this period? If not, why not?*

4. Compound and weighted variables

Introduction

Often the data which is given in a table or graph is not an original, single data series, but a combination of such series.

The section above on real and nominal variables is one particular example. You will come across many other examples. For instance,it is often necessary to weight by the size and/or the structure of the population. Some examples are given below:

(i) The unemployment rate is the number unemployed divided by the working population.

(ii) A similar, but more complex weighting is necessary to compare death rates between areas. The crude death rate is the death rate per head of population, but this takes no account of the age and sex distribution of the population. Standardised mortality rates adjust the crude death rate for these factors.

(iii) Another complex weighting is required to compare the standard of living between countries. GNP figures by themselves are not applicable even as a starting point for such a comparison, the figures need to be weighted by population to give GNP per head, and the resulting figures translated into a common currency for comparison. This later step is very difficult, because although exchange rates are readily available for many countries on a daily basis, these fluctuate considerably and often do not reflect underlying price differences between the countries. Statisticians produce series trying to adjust for these problems.

Interpretation

Before interpreting the data it is necessary to appreciate the constituents of such a complex data set. Consider the definitions of the variables involved and then why the data has been combined and presented in this manner. In presenting data yourself, such data combinations will often illustrate an argument better than presenting all the raw data for the reader to assess.

Exercise 8

The table below gives the savings ratio for the U.K. in the 1980s. This is the ratio of savings to incomes.

(i) *consider the trends in the table*

(ii) *given that what is not saved is consumed, do the figures give credence to the view that the boom in the U.K. economy 1986-89 was a consumption-led boom?*

(iii) *does the series support the view of a stable consumption function over the period (consider the findings on consumption and income discussed on page 19). Does the presentation of data in this form make any of the trends clearer?*

Savings ratio, U.K.

Year	%	Year	%
1980	13.4	1987	7.1
1981	12.6	1988	7.0
1982	11.3	1989	7.2
1983	9.7	1990	8.6
1984	11.1	1991	10.1
1985	10.7	1992	12.3
1986	8.7	1993	11.5

Source: Economic Trends

SUMMARY OF KEY POINTS

1. Data transformations are sometimes necessary, and often useful, in presenting data to argue a case.
2. Data is often given in percentages rather than levels. With cross sectional materials, percentages provide an easy way to compare the importance of various categories. With time series data the percentage change gives a simple growth/decline rate over the period.

3. Indices measure the change in a variable from a base point and are useful in comparing series over time. Indices are a common method of presenting the average change of a group of related items, for instance the prices of goods in the Retail Price Index.

4. Data in "real terms" has been adjusted for inflation. If the concern of an hypothesis is with quantities (as is commonly the case), it is "real" data that needs to be examined.

5. It is often useful to combine data sets to illustrate a case. This often involves "standardisation" in some respect.

STUDY D

1. Convert the data on private housebuilding completions in Table D1 into percentages (or proportions) of the total number of houses in each size group.

2. Using the data on Real Gross Domestic Product in Table D2, work out (a) annual growth rates over a five year period, (b) the average compound growth rate over the same period. Compare the average of the annual growth rates with the compound figures. Finally convert the data to index numbers with 1980 as the base of 100, and graph the resulting series.

3. Compare the trends of nominal and real expenditure on drink and tobacco in Table D3. It may be useful to convert them to index numbers with 1980=100 before graphing. Do the two graphs tell the same story?

4. Divide the data on Real GDP in Table D2 by the population data on the same countries in Table D4 to create a new compound variable: Real GDP per capita. Comment on the differences in living standards between the countries.

V ANALYSING AND PRESENTING DATA USING STATISTICAL DESCRIPTION

Statistical description uses formulae to describe some attribute of the data. The aim is to provide a summary of the data that provides important information relating to the investigation in hand – an overall measure. A concise numerical description is often preferable to long tables of figures or even graphs. The information given by these formulae helps in the interpretation of the data, particularly where the values are variable and comparisons are required. The formulae considered here include the mean, the standard deviation and correlation.

The word ***description*** here is important. Statistical formulae are used to provide overall (or aggregate) measures that aid the description of the data. Some of these statistics can be used in formally testing hypotheses – statistical inference – but this development is not covered here. This does not imply that these statistics cannot be used to support the general investigation of an hypothesis, nor provide information useful in arguing a case. Indeed they are and should be so used.

In the sections that follow the various measures are defined in words and then a formula is given. The formula may look complicated. Do not spend a long time trying to understand the mathematical notation if this is not familiar to you, but concentrate on the example which lays out carefully the steps in the calculation. There is nothing more difficult than division and squaring a number in the actual arithmetic. Calculators and computer programs can be used to do the more complicated arithmetic. The interpretative aspects are stressed here and the situations in which such statistical descriptions are useful.

1. The middle value

Often of greatest concern is to measure the middle, or centre, of the data – the 'norm'. Three measures are commonly used:

The arithmetic mean

This is the most widely used measure of the middle, and the only one that has a standard mathematical formula that can be further developed.

■ **definition**

The average of the data, that is the total of the data divided by the number of items.

example

Table 11 gives the rate of unemployment in GB over the period 1975(Q1) to 1979(Q4). In order find the mean unemployment rate for 1975, we sum column 2, that is

$$2.2 + 2.6 + 3.1 + 3.6 = 11.5,$$

and divide by 4, which gives a mean of 2.9 to one decimal place.

The mean for the whole period 1975 to 1979 is obtained by summing all the observations in the table (given the way this table is set out this requires summing columns two to six) and dividing by the number of observations. Here the overall sum is 78.2, and the number of observations is 20, giving:

$$\text{Mean Unemployment Rate (1975/79)} = \frac{78.2}{20} = 3.91$$

Table 11: Unemployment Rate % GB.

	1975	1976	1977	1978	1979
Q1	2.2	3.9	4.2	4.4	4.0
Q2	2.6	4.1	4.2	4.3	4.0
Q3	3.1	4.2	4.4	4.3	3.9
Q4	3.6	4.2	4.5	4.2	3.9

Source: Economic Trends

formula

$$\overline{X} = \frac{\Sigma X_i}{n}$$

where $\overline{X}$ = the mean of X_i
X_i = the sum of the observations
n = number of observations
Σ = sum of

use and limitations

The arithmetic mean is the predominant measure of central tendency that is used. It is the best understood 'average' and is relatively easy to calculate. Every value in the series is taken into account. It is a very useful statistic in making quick comparison between different time periods, countries etc.

A limitation with some data occurs if there a few extreme values at one end. These values will drag the mean towards them and this may make the mean unrepresentative.

weighted average

An important and widely used extension of the average is the weighted average. This is used if the data is grouped.

Table 12 gives the percentage of households by size. In order to find the average size of household we need to multiply each possible household size by the percentage in that category, that is the frequency. This is illustrated in column 3, labelled f_iX_i (Household size is X_i and the % is the f_i). The average is the sum of this column divided by the total frequency (which in this case 100 since the table is in percentages). This gives:

244/100 = 2.44.

Table 12: Household size, %, 1991, GB

Household Size	%	f_iX_i
1 person	27	27
2 person	34	68
3 person	16	48
4 person	16	64
5 person	5	25
6 person	2	12
Σ		**244**

The formula for the weighted average is:

$$\bar{X} = \frac{\Sigma f_i X_i}{n} \quad \text{where } f_i = \text{the frequency of } X_i$$

The median

definition

The median is the central observation. To find the median the data needs to be ordered and then the middle figure found.

use

The median is a better measure of the 'centre' to use than the mean in circumstances where a few outliers in one direction pull the mean in that direction.

examples

In the example above (Table 11) the median can be easily found from the ranked data. In Table 13 the unemployment percentages have been put in ascending order. There are two 'middle' values, 4.1 and 4.2 (shown in bold), and the median is the average of these, i.e. 4.15.

However, for this data the mean is probably the better central measure, as there are no outliers.

Table 13: Ranked unemployment rates, 1975/9

2.2
2.6
3.1
3.6
3.9
3.9
3.9
4.0
4.0
4.1
4.2
4.2
4.2
4.2
4.2
4.3
4.3
4.4
4.4
4.5

With the household size data in Table 12, the median can be read directly as 2, since the categories 1 person and 2 person households account for 61% of the total. The median is giving some extra information, since there are a few households with a large number of members which increases the mean.

Mode

- **definition**

The mode is the most common result in a set of data. Thus to identify the mode in a set of data the responses need to be collated into frequency groups. The most common response is the mode.

- **example**

With the household size data the mode can be directly read off from the table. The mode is the largest group, in this case a household size of 2.

Finding the mode for the unemployment data in Table 13 requires a little more inspection. Which unemployment rate is the most common? With this set of data the answer is 4.2%, as there are five observations of this magnitude over the period and no other value has a greater number. If visual inspection becomes too complicated, to find the mode order the data and then work out the frequency of responses.

In these examples there was only one mode, but in some cases there may be several responses with the same frequency.

- **use and limitations**

The mode is easily understood and easily identified. As with the median it is not affected by a few unduly high or low values, but it does not take account of the range of the data responses. If you need to find the most common outcome, the mode is the correct statistic to quote.

Exercise 9

Using the data in the table below:

(a) *Calculate the mean of unemployment over the period 1985(Q1) to 1989(Q4). Compare this to the situation in the second half of the 1970s.*

(b) *Calculate the median and mode of this data. Why is the arithmetic mean a better measure of central tendency than the median and mode for this data?*

Unemployment rate, % GB, 1985/89

	1985	1986	1987	1988	1989
Q1	10.7	10.9	10.7	8.7	6.7
Q2	10.8	11.1	10.3	8.3	6.3
Q3	10.8	11.1	9.8	7.7	6.0
Q4	10.9	11.0	9.2	7.3	5.6

Source: Economic Trends

2. Measures of dispersion

Often, as well as the average value, we want to measure the variability of a set of data. Two sets of data may have the same mean, but very different dispersions.

The range

- **definition**

The range is simply the difference between the lowest and highest observation. In stating the range sometimes the highest and lowest values are given, sometimes the difference.

- **example**

The unemployment rates for 1975–79 are repeated below in Table 14. The range of the data is 2.2–4.5. These are highlighted in bold. The difference between the highest and lowest unemployment rates is thus 2.3.

- **uses and limitations**

This is the simplest measure of dispersion to establish, but it takes no account of the distribution of the observations in between the two ends. There may be just one outlying observation that is increasing the range.

The Mean Absolute Deviation (MAD)

- **definition**

In considering differences from the mean value, some observations will be above and some below the mean. The differences must sum to zero. The mean absolute deviation takes account of these differences by ignoring the sign. It measures the absolute deviation from the mean over all observations.

example

In Table 14, column 3 gives the difference between the observed value and the mean. Column 4 gives the values ignoring the sign and summing these gives the numerator in the formula. Dividing by the number of observations, in this case 20, gives the value of the mean absolute deviation.

MAD = 8.34/20 = 0.417

A larger value implies a larger dispersion.

formula

$$MAD = \frac{\Sigma|X_i - \bar{X}|}{n}$$

where the symbols have the same meaning as before, with the addition of | | which means take the modular, or absolute value, i.e. ignore the sign.

uses and limitations

The mean absolute deviation is relatively easy to understand and calculate. Account is taken of all the observations in the measure. It is a good measure to use when there is no reason to treat large deviations from the mean in any different manner to small ones.

The Standard Deviation

definition

The standard deviation is another measure that takes into account all the observations. The negative and positive differences from the mean in the formulation are taken account of by squaring the differences. The variance measures the average squared deviations from the mean. The standard deviation is the square root of this.

example

In the example in Table 14, column 3 gives the difference between the observed value and the mean. Column 5 gives this value squared, and the bottom of this column gives the sum of these values. This is divided by the number of observations (20) and the square root is calculated to give the standard deviation

$s^2 = 6.996/20 = 0.59$ $\quad$ $s = 0.78$

Again a larger value implies a larger deviation. The units of measurement of the standard deviation are the same as for the original data, thus in this example the standard deviation is 0.78%.

Table 14: Unemployment rates in GB.

			$X-\bar{X}$	$\lvert X-\bar{X}\rvert$	$(X-\bar{X})^2$
1975	**Q1**	**2.2**	-1.71	1.71	2.924
	Q2	2.6	-1.31	1.31	1.716
	Q3	3.1	-0.81	0.81	0.656
	Q4	3.6	-0.31	0.31	0.096
1976	**Q1**	3.9	-0.01	0.01	0
	Q2	4.1	+0.19	0.19	0.036
	Q3	4.2	+0.29	0.29	0.084
	Q4	4.2	+0.29	0.29	0.084
1977	**Q1**	4.2	+0.29	0.29	0.084
	Q2	4.2	+0.29	0.29	0.084
	Q3	4.4	+0.49	0.49	0.240
	Q4	**4.5**	+0.59	0.59	0.348
1978	**Q1**	4.4	+0.49	0.49	0.240
	Q2	4.3	+0.39	0.39	0.152
	Q3	4.3	+0.39	0.39	0.152
	Q4	4.2	+0.29	0.29	0.084
1979	**Q1**	4.0	+0.09	0.09	0.008
	Q2	4.0	+0.09	0.09	0.008
	Q3	3.9	-0.01	0.01	0
	Q4	3.9	-0.01	0.01	0
	Σ			**8.34**	**6.996**

- **formula**

$$s = \sqrt{\frac{\Sigma(X-\bar{X})^2}{n}}$$

where s = standard deviation and the other symbols have the same meaning as above.

uses and limitations

The standard deviation is the standard statistical measure of deviation. It gives greater weight to larger deviations than to smaller ones in the calculation since the differences are squared. It is a common assumption that large deviations are of relatively more importance. It is the most frequently used measure of dispersion and one that is used in probability statements when more formal statistical testing is employed.

Exercise 10

Using the data on GB unemployment, 1985-89 (given for Exercise 9) calculate the range, mean absolute deviation and the standard deviation. Compare this with the situation in the second half of the 1970s described above.

3. Correlation

description

The correlation coefficient is a measure of the strength of the linear relationship between the variables.

scatter diagrams

Graphical illustration using scatter diagrams is a useful introduction to considering relationships. A scatter diagram or XY plot illustrates the relationship between two variables, X plotted on the horizontal axis and Y plotted on the vertical axis. The association may be positive, negative or non existent. It is used in studying the association between variables, and the points are not usually connected as in a line graph.

The axes are plotted as in a line graph and the data points are mapped into the XY space. The points are not joined together. In the example shown in Graph 12, personal income (after adjusting for taxes and inflation) is on the X axis and car production on the Y axis. The graph plots the data pairs over the period 1980–1990 for the UK.

Graph 12 : Passenger car production and Real Personal Disposable Income, 1980/90

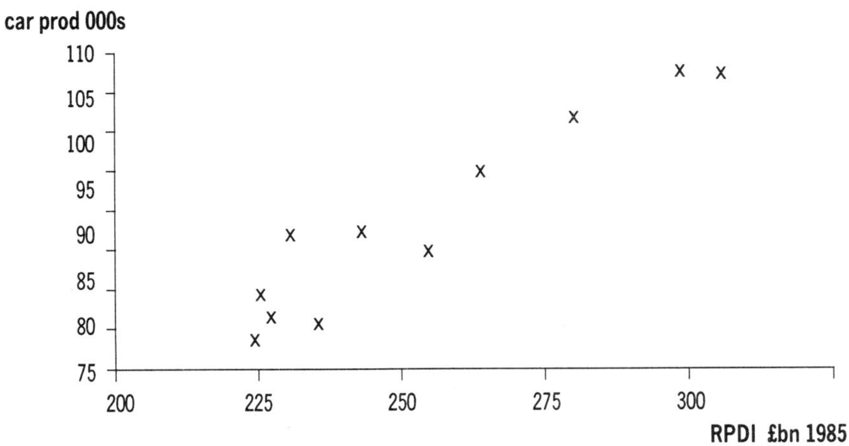

Source: Economic Trends

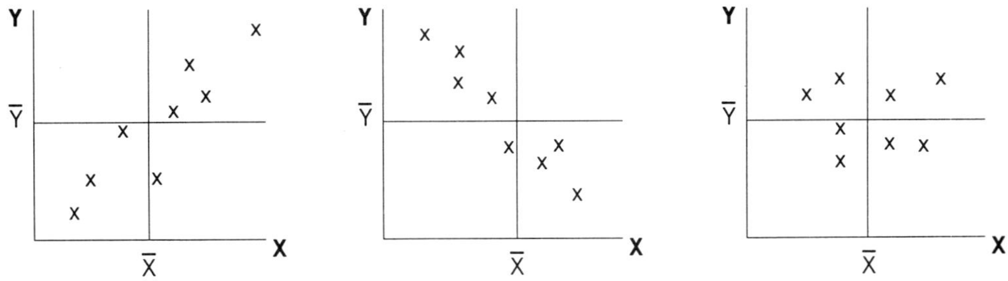

(a) positive association *(b) negative association* *(c) little association*

The diagrams above illustrate in general terms the various possibilities. The mean of the Y and X variables are given, along with the data points. (a) shows a positive association and (b) a negative association. If the data is scattered there is little to no association, as shown in (c).

the correlation coefficient

The correlation coefficient formally measures the strength of the association. The formula is designed so that the correlation coefficient lies between -1 and +1.

The closer the correlation coefficient to +1, the higher the positive association. The two variables increase and decrease at the same time, as in diagram (a) above. This is referred to as a direct relationship.

The closer the correlation coefficient is to -1, the higher the negative association, as in diagram (b) above. The two variables tend to move in opposite directions. This is referred to as an inverse relationship.

The closer to 0 is the correlation coefficient, the less the degree of association between the variables. Little association is illustrated in diagram (c).

There are several points to remember:

(i) The above construction and the formula below is for the **Pearson** or **product moment correlation coefficient** which can be used for quantitative data and gives a linear association.

(ii) A low value of the correlation coefficient does not imply that there is no association, only that there is not a simple linear association. The association may be non linear, or may be complicated by other factors that are also affecting the variables.

(iii) A negative sign on a correlation coefficient, say -0.9, does not imply a weak association, but a strong negative correlation.

Calculation

The formula for the product moment correlation coefficient

looks a little complicated, but is easy to apply and is usually worked out by computer.

$$r = \frac{n\Sigma XY - \Sigma X \Sigma Y}{\sqrt{[n\Sigma X^2 - (\Sigma X)^2][n\Sigma Y^2 - (\Sigma Y)^2]}}$$

Below is an example using just six data points for illustration. It uses imaginary data to give simple calculations, but consider that X stands for unemployment and Y for inflation in some country.

Table 15: Correlation Calculations

	X	Y	X^2	XY	Y^2
	5	25	25	125	625
	8	31	64	284	961
	2	8	4	16	64
	7	20	49	140	400
	3	22	9	66	484
	12	19	144	228	361
Σ	**37**	**125**	**295**	**859**	**2895**

$$r = \frac{6 \text{ x } 859 - 37 \text{ x } 125}{\sqrt{\{[6 \text{ x } 295 - 1369][6 \text{ x } 2895 - 15625]\}}}$$

$$= \frac{5154 - 4625}{\sqrt{(401 \text{ x } 1745)}}$$

$$= \frac{529}{836.5} = 0.63$$

This shows that there is a moderate degree of positive association between X and Y, that is between unemployment and inflation for the imaginary country.

Exercise 11

What does the correlation coefficient show in these cases?

(a) r = 0.8 between consumption and income over the period 1970-1990.
(b) r = 0.3 between rainfall and tons of potatoes harvested per acre.
(c) r = -0.15 between the crime rate and the temperature on a particular day.

(d) r = -0.95 between the price of strawberries and the quantity sold. Are these correlations what you would expect?

4. Finding the trend

In times series often the key question is to distinguish the trend. Important questions include:

- is the general movement up, down or stationary?
- what is the magnitude of the change, on average, over a particular period such as a year?
- has the trend remained unchanged over the whole of the period of the data, or has it changed direction or size?

It can be quite difficult to distinguish the trend, and changes in it, from the raw data.

Table 16: New car registration, GB, 1989/92

	'000s
1989 Q1	836
Q2	748
Q3	964
Q4	520
1990 Q1	772
Q2	636
Q3	844
Q4	426
1991 Q1	611
Q2	452
Q3	712
Q4	359
1992 Q1	539
Q2	481
Q3	709
Q4	404

Source: Economic Trends

This table gives sales of cars in the UK 1989/92.It is difficult to detect the trend and properly analyse the figures in this form. The data is highly seasonal and this tends to obscure the other changes.

Freehand curve fitting

There are various method of estimating the trend. Perhaps the most straightforward is freehand curve fitting. This method is limited by its subjectivity; it is not easy to judge precisely the trend and even more difficult to say at what point a particular trend has changed. Graph 13 gives the new car data in graphical form, with a freehand line fitted to try and identify the trend.

Graph 13: New car registration, GB, 1989/92

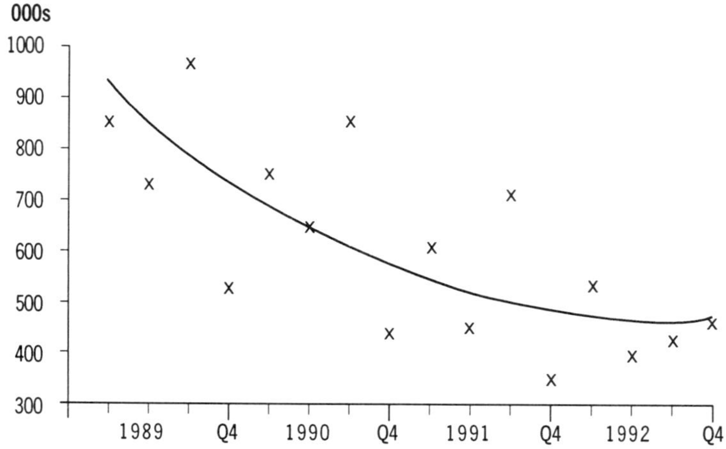

It is possible to trend fit using statistical techniques – for a linear (straight line) trend, the regression technique covered in Chapter VI gives a good estimate. However, this is only appropriate if the trend is approximately linear – here for instance the last two or three observations are not showing quite the same pattern as previous quarters. Has there been a change of trend, or will the downward trend reassert itself? There is no correct answer here, but there are other techniques that can be used, and which is best is a matter of judgement given the particular data. Two fairly straightforward techniques are briefly considered below. Other, more complicated techniques also exist.

Moving averages

A moving average smoothes a series, removing short run fluctuations so that the underlying movement in a series can be distinguished.

The first thing to decide is the length of the averaging process. A longer length has the advantage of smoothing the series more, it has the disadvantage of not quickly picking up changes in the underlying trend. The length is therefore a matter of judgement. With the new car data being used as an example, the length needs to be enough to remove the seasonality, at least four periods. Once you have decided on the length of averaging period, the calculation of the moving average is straightforward:

- average the first four observations. This gives the moving average for the mid point. In the example (836+748+964+520)/4 = 766.5
- to obtain the next period moving average, drop the first period's data, add the next, and average these new four observations, i.e. (748+964+520+772)/4 = 751

This is shown in Table 17. Forecasting the trend from a moving average is a matter of judgement. Two possible options are to take the last value available of the moving average, or a freehand trend of recent moving averages can be taken. Graphing the moving average is often useful in distinguishing the trend.

Table 17: New car registrations, four period moving average

Year	Calculation	Moving Average
1989 Q1	—-	
Q2	—-	
Q3	(836+748+964+520)/4	767
Q4	(748+964+520+772)/4	751
1990 Q1	(964+520+772+636)/4	723
Q2	(520+772+636+844)/4	693
Q3	(772+636+844+426)/4	670
Q4	(636+844+426+611)/4	629
1991 Q1	*	*
Q2	(426+611+452+712)/4	550
Q3	(611+452+712+359)/4	534
Q3	(452+712+359+539)/4	516
1992 Q1	(712+359+539+481)/4	523
Q2	*	*
Q3	(539+481+709+404)/4	533
Q4	——	

To be fully accurate, the moving averages in the table above apply to the midpoint between the quarter it is in and the quarter above.

Exercise 12

*Table 17 gives the moving average for the new car data, but two calculations are missing. These are shown by *. Calculate these moving averages* *to complete the table and then graph your results. Compare the graph to that of the original data. Comment on the trend.*

Exponential smoothing

A popular variation of the moving average is exponential smoothing. This still averages the data to smooth it, but more recent data is weighted more heavily in the calculation. This has the advantage that stress is placed on the more up-to-data situation. However, a drawback is that if too much stress is placed on very recent observations the smoothing process is reduced. The exponential moving average is a little more complicated to calculate and is not covered here, but many popular computer programs will do this automatically.

In calculating an exponential moving average the weight given to the final observation has to be chosen. The weights for the other periods then decline exponentially. If the weight given to the last observation is 0.5, the weight of the previous observation is 0.5 x 0.5 (=0.25), and the one before that is 0.5 x 0.5 x 0.5 (=0.125). The higher the weight for the first period, the lower the degree of smoothing.

Graph 14: Exponential smoothing of production statistics

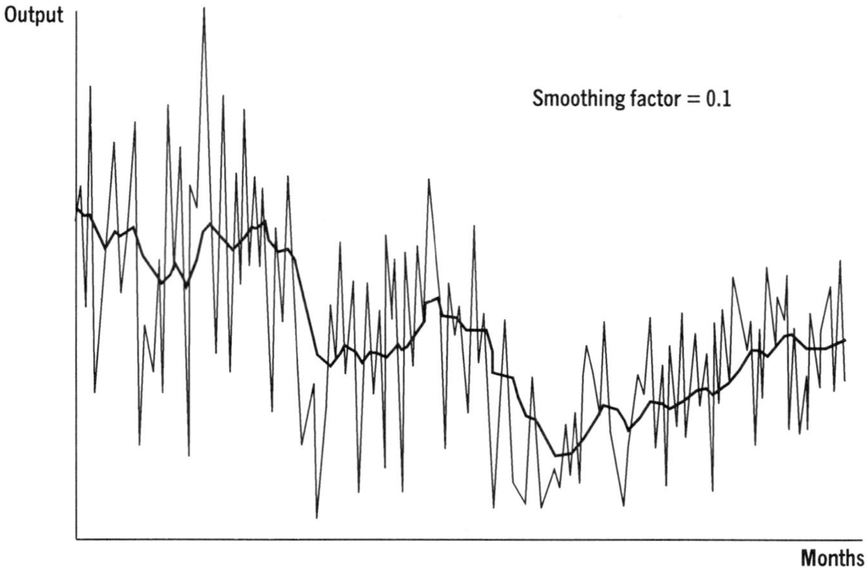

An example of exponential smoothing is given in Graph 14, which shows a highly variable series of monthly output data. The smoothing factor chosen, 0.1, is actually quite large, but given the variability in the data the resulting trend line is still quite variable. Nevertheless the exponentially smoothed line, shown as the dark line, strongly dampens the oscillations.

SUMMARY OF KEY POINTS

1. Statistical description provides concise figures to summarise important characteristics of a data set.
2. The most common measure of the centre of the data is the mean. The median and the mode may be useful in certain circumstances.
3. The spread, or dispersion, of the data is also important to consider in many circumstances. The standard deviation is the most important measure of this, but the range and mean absolute deviation can also be used.
4. The strength of association between variables can be measured (for linear association) by the correlation co-efficient. The association may be positive or negative.
5. Identifying the trend is often the most important aspect of interpreting time series data. There are many methods for doing this. Those presented, including freehand fitting, moving averages and exponential smoothing, are relatively straight forward but useful approaches.

STUDY E

1. Using Tables E1 and E2, calculate the:

(a) mean, median and mode
(b) range, mean average deviation and standard deviation of life expectancy in Africa and Europe.

Use your results to write a short comparison of life expectancy between the two continents.

2. Construct :

(a) a moving average
(b) an exponential moving average

for energy consumption using the data in Table E3. Comment on the results of the different averaging processes in this case, and on the trend in the series.

VI LOOKING FOR CAUSAL RELATIONS

1. Direction of causation

In interpreting data a common approach is to consider the relationship, or association between two or more sets of data. For instance, the relationship between consumption and income or the association between rainfall and crop acreage. The question of causation is important to consider. If two data items A and B vary together there are four possible cases:

- A and B are unrelated – the fact that the series move together is just pure coincidence
- A is causing the changes to B
- B is causing the changes to A
- A and B are jointly determined in a system with other variables

With the statistical investigation of relationships, the starting point should be an hypothesis which generally will come from the material covered in your course. This should mean that unrelated data will not be examined in the search for causal relationships. It makes no sense to investigate associations between data series where no rationale can be found for a link. For instance, it may be possible to order the heights of the Himalayan peaks in such a way as to give a strong statistical relationship with changing unemployment in the UK in the 1980s. However, there is no possible link between the two and such a calculation would be a waste of time and effort. Indeed, it would be a misuse of statistics.

The direction of causation, whether A is causing B or B causing A is also often a matter for theory. If there are no time lags involved, statistics cannot establish the direction of causation, it is purely a theoretical matter. However, theory is not always helpful: there may be competing theories that suggest different directions of causation. A well known dispute of this kind concerns inflation and the money supply. These variables are positively related (at least for some definitions of the money supply over some periods), but does an increase in the money supply cause inflation, or does the money supply rise because of the inflation? Monetarist economists accept the first causation, post-Keynesian economists believe it is the second that applies. Statistics may help if there are lags involved in the process. Most would accept that if a change in A occurs first, it is A causing B, and vice-versa. However, there are many difficulties with such interpretation, some of which are covered in the latter part of this section.

If A and B are jointly determined, it implies that there is a simultaneous relationship. A and B are two variables in a larger system. Simultaneity is widespread in economic and business relationships. For instance, at an individual level it may be accepted that a change in income levels leads to a change in consumption. At the aggregate level of national income and consumption this is also true, but an increase in consumption will also increase income, as sales and production are increased.

It is also often the case that more than one variable is involved. A may cause changes to B, but so may items C, D and E. For instance, income changes may cause consumption changes, but consumption may also be affected by interest rates, changes in wealth, etc. This sometimes brings difficulties when examining data for causal relationships using simple techniques, such as graphs. Relationships may not show up because 'other things are not equal', i.e. many of the variables are changing at the same time.

The sections that follow cover the descriptive use of graphs and statistical formulae to quantify associations and relationships. The statistics covered in this section can be developed in order formally to test hypotheses using statistical inference. However, this development is not covered here.

2. Use of graphs

The first step when examining relationships is to become familiar with the data. The graphical techniques described in earlier sections are very useful here. Simple line graphs, bar charts, etc. can help in gaining an appreciation of trends and associations. Graphs of two related variables, as considered in Chapter II 2(d) can also be used.

The scatter diagram, examined in Chapter IV.3, is particularly useful in ascertaining relationships. If, in the hypothesis being considered, it is possible to distinguish causation, the variable causing change is known as the independent variable, the variable being affected is known as the dependent variable. When plotting these in a diagram or graph, generally the independent variable is placed on the X axis and the dependent on the Y axis. A line of 'best fit' may be shown, which could be a simple straight line fitted by eye, or the line could be the result of mathematical calculations.

The example below in Graph 15 repeats the data in Graph 12. Since the hypothesis is that changes in personal income affect car production, personal income (after adjusting for taxes and inflation) is on the X axis and car production on the Y axis. The graph plots the data pairs over the period 1980-1990 for the UK.

In the social science and business areas exact relationships are unlikely. The lack of experimental conditions means that a number of other, perhaps small, factors that are affecting the situation cannot be kept constant. In the example, as income has increased so has the production of passenger cars. The relationship is likely to come about because as people become better off, they demand more cars. However, the relationship is far from exact, as can be seen by the deviations from the line.

Graph 15: Passenger car production and Real Personal Disposable Income, 1980/90

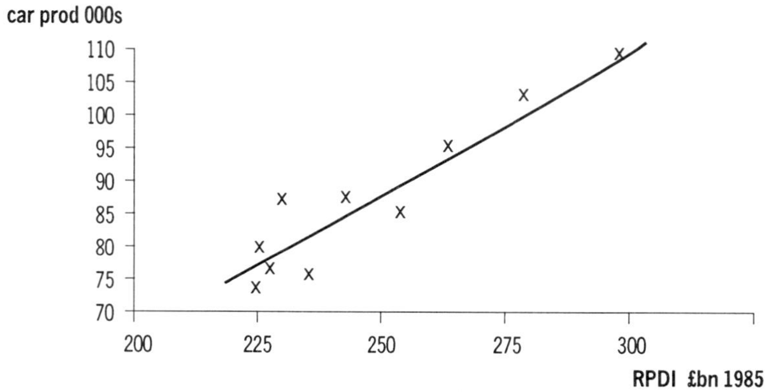

Source: Economic Trends

The line of best fit gives an estimate of the relationship between the two variables. In the example, the line of best fit has been drawn by hand. It indicates a positive relationship. The slope of the line suggests that for every 10 billion pound increase in incomes there will be approximately a 8000 increase in the production of cars.

It is important to remember that "Graphs cannot prove anything". They illustrate possible relationships. The interpretation of a graph should be in terms of the hypotheses of the subject under consideration. One particular point arises when interpreting scatter graphs. If no association seems to exist between two variables on such a graph it does not necessarily imply that there is no correlation. It may be that a third factor is having a great effect on the variables and this is complicating the picture. This is an outcome of the problem noted above, that in the social science and business areas experiments cannot be conducted to keep other things constant.

Exercise 13

As GDP grows, so imports are expected to grow. Construct a scatter diagram from the data given below showing imports and GDP in the U.K. The data has been adjusted to remove the effect of inflation. Does the graph show a positive, negative, or zero relationship? Does the scatter diagram support the hypothesis above? (This is the same data as in Graph 3, but you are being asked to present it in another way. Either way is equally valid).

Percentage change in imports and Real GDP, 1983/93

Year	Imports % change	GDP % change
1983	6.2	3.6
1984	9.0	2.0
1985	2.5	3.7
1986	6.4	3.9
1987	7.2	4.4
1988	11.2	4.7
1989	6.9	2.2
1990	0.5	0.56
1991	-5.7	-0.24
1992	5.9	-0.4
1993	2.6	1.9

Source: Economic Trends

3. The line of best fit

Description

The use of the ruler and eye to determine the 'line of best fit' is highly subjective. What is required is a more formal method of estimating the line.

The mathematical formula for a straight line is:

$Y = a + bX$, where Y = dependent variable
X = independent variable
a = intercept
b = slope

Diagrammatically:

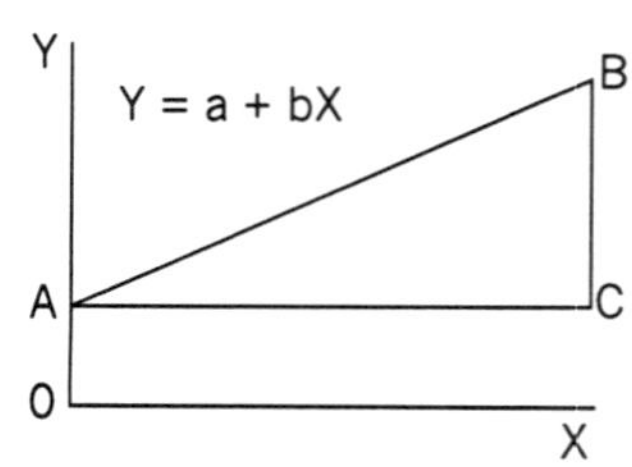

The regression line gives an estimate of a and b, where a = 0A and b = BC/AC.

Construction

How is the equation to be estimated? The deviations of the data points from the line must be small for a good fit. Mathematically, any deviation above the line is positive and any deviation below the line is negative. For a good fit the deviations must sum to zero. However, this is not sufficient as a criterion to obtain an estimate of the line – there are many lines that satisfy this criterion. Thus a second criterion is also employed, that of least squares. The equation is estimated so that the sum of the squared deviations from the line are minimised. The diagram left illustrates this. 0_1 to 0_6 are the data points with the vertical deviations around the regression line shown. The regression line is chosen so that the sum of the deviations is equal to zero, and the sum of the squared deviations is as small as possible. The least squares criterion weights large deviations from the line more than small deviations.

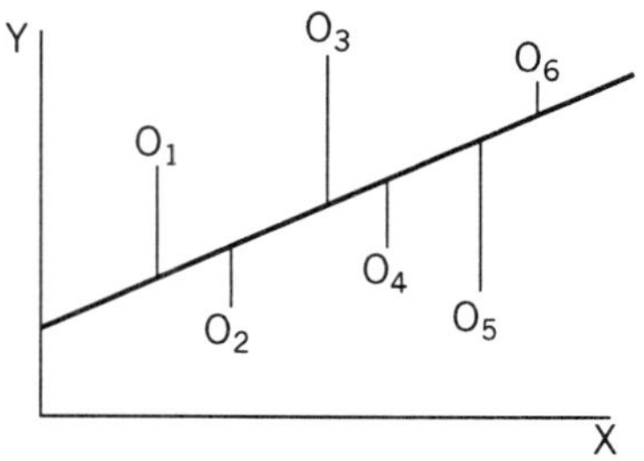

Usually these days the actual calculation is done by computer, and certainly understanding the principles and being able to interpret the results is far more important than being able to do the calculation. The calculation for the example below is given at the end of this section for those who wish to pursue it.

Interpretation

A regression line is estimated as:

$C = 100 + 0.8Y$ where $Y =$ Personal Disposable Income
$C =$ Consumption
both in £mns at 1990 prices

This says that on average when Y goes up by one unit C goes up by 0.8 units, and if $Y = 0$ then $C = 100$.

To pursue the interpretation further requires a consideration of the units that C and Y are expressed in. Here they are in £ million at 1990 prices, so that for every 1 million increase in real personal disposable income, consumption increases by 0.8 million. The coefficient of Y (i.e. 0.8) is of particular importance in economics – it is known as the marginal propensity to consume.

The units are of importance. If for instance, the income was expressed in billions rather than millions, the coefficient value would be 800 instead of 0.8, i.e. a 1000 times greater.

This point is of particular importance if the estimated line includes more than one independent variable, which is known as multiple regression. For instance,

$C = a + 0.8Y - 350R$ where R = the rate of interest written as a proportion.

It may appear at first glance that the interest rate is more important than income in determining consumption, because the size of the coefficient is greater (350 compared to 0.8) but this is not necessarily so, it depends on the size of the units. Here Y is measured in millions and the interest rate is measured as a proportion (.1 is a 10% interest rate). To establish the more important variable requires a consideration of the likely range of values that can be taken by Y and R. Given the size of the income variable in the UK, in this case income is the more important determinant.

The interpretation of the intercept in terms of the hypothesis is often far less important than the slope co-efficient. Usually the possibility of the independent variable actually taking the value zero is non-existent, but the intercept is important mathematically because otherwise the equation is constrained to go through the origin. Here it is also important in determining the average propensity to consume, which is C/Y or (a + bY)/Y.

Forecasting

An estimated regression line can be used for forecasting, if the values of the independent variables are known or can be estimated. If, in the example above, Y is expected to be equal to 100 and the rate of interest to 0.1, the forecast value for C is :

$$C = 100 + 0.8(100) - 350(0.1)$$
$$= 100 + 80 - 35 = 145$$

Calculation

Table 18 gives the steps in calculating a regression line using a small number of observations to illustrate the method. The data used is the same imaginary data used in the correlation calculation in Chapter V.3. X is unemployment and Y is inflation.

The formula for the slope is:

$$b = \frac{n\Sigma XY - \Sigma X \Sigma Y}{n\Sigma X^2 - (\Sigma X)^2} = \frac{6 \times 823 - 37 \times 125}{6 \times 297 - (37)^2} = 0.78$$

The formula for the intercept is:

$$a = \frac{\Sigma Y - b\Sigma X}{n} = \frac{125 - 0.78 \times 37}{6} = 16.02$$

The estimated line is therefore $Y = 16.02 + 0.78\,X$

Table 18: Regression calculations

	X	Y	X^2	XY
	5	25	25	125
	8	31	64	248
	2	8	4	16
	7	20	49	140
	3	22	9	66
	12	19	144	228
Σ	**37**	**125**	**295**	**823**

Exercise 14

A demand curve is estimated to be: $Q = 10.7 - 0.6P + 0.4Y$
where Q = *quantity of widgets demanded, in 000s.*
P = *price of widgets in £s*
Y = *real personal disposable income, 1990 prices, £ bns*

(i) *interpret the results*

(ii) *do the results support the law of demand? (that lower prices increase the demand for goods)*

(iii) *if the price is expected to be £10 and income £68bns, what is the quantity of widgets demanded?*

4. Lagged relationships

Example

When examining two or more sets of time series data, we are usually considering relationships. This may be straightforward, in that as one series increases so does the other (a positive relationship) or as one series increases the other decreases (a negative relationship).

However, many relationships are not that straightforward. One important possibility to consider is that of a lagged relationship. This is where one variable is affecting another, but with some time delay. This may be of a fairly short period, of say a quarter, or a longer delayed effect of perhaps 18 months to two years.

Graph 16(a): Changes in the money supply and inflation, UK, 1973/81

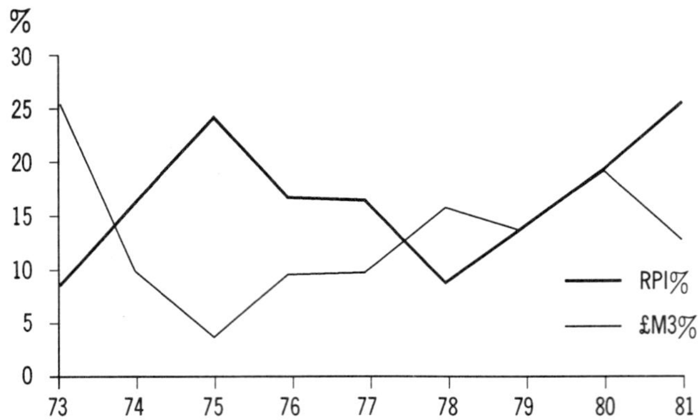

Source: Economic Trends

Consider Graph 16(a) showing the rate of growth of the money stock (by the definition £M3) and inflation in the 1970s. At first glance it may be thought that inflation and the money supply appear negatively correlated. However, the inflation rate can be interpreted as following the increases and decreases in the money supply with around a two year lag. The Graph 16(b) shows the inflation lagged two periods to illustrate this. The fit is not exact, but the large growth in the money supply in 1972/73 and again in 1978/79 was followed by similar increases in the inflation rate.

Graph 16(b): Changes in the money supply lagged by two years and inflation, UK, 1973/81

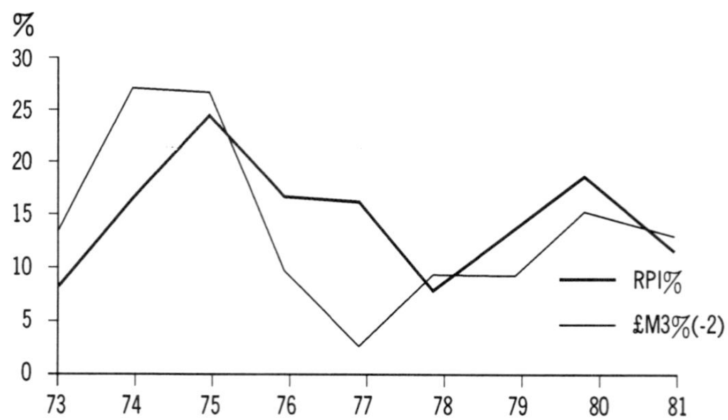

Interpretation

In considering such a possibility it is necessary to ask whether such a lag is reasonable. Does the theory suggest such a delayed response? There is a need to be very careful. We could use lags to justify nearly any association. There does need to be an appeal to theory, as well as to the data, to establish a justification for such a delayed response. In the example, the appeal could be to the monetarist theory of inflation that suggested such a delayed response.

However, this is not sufficient. Graph 17 of the 1980s shows the relationship did not hold over this period. There appears to be little link between the rate of growth of the money supply (as least as measured by £M3) and inflation, even if various possible lags are considered over this later period. This would suggest that the hypothesis needs either amending or discarding. However, it should be remembered that the hypothesis needs formally testing using the statistical evidence. Graphical interpretation can suggest only that an investigation is worth pursuing.

Graph 17: Changes in the money supply lagged by two years and inflation, UK, 1980/88

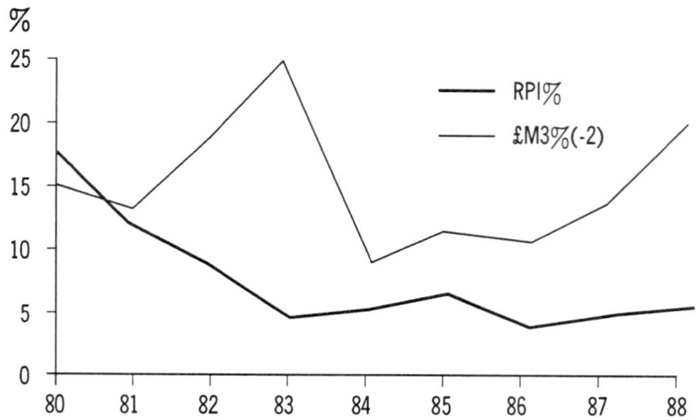

It is also important to recognise that links between variables will not always show up in a table or in a graph, however hard we look. This may be the case if a variable is being affected by two or more factors. To establish such relationships more advanced statistical techniques are required which allow you to estimate the effect of a particular variable while keeping other factors constant.

Exercise 15

The table below gives the percentage changes in personal incomes (after adjusting for inflation and taxes) and real GNP (that is total output in the economy).

(i) *graph the percentage change in personal incomes and real GNP over time.*

(ii) *redo the graph with the percentage change in personal incomes lagged one period compared to GNP.*

(iii) *does your graph support the idea that there might be a lagged relationship between the two variables? Can you think of any reasons why changes in incomes might follow changes in GNP?*

Real Personal Incomes and Real GNP, % change, 1981/90

Year	Incomes % change	GNP % change
1981	0.9	-0.9
1982	-0.4	1.2
1983	2.7	4.0
1984	2.6	1.9
1985	2.5	3.8
1986	5.0	3.4
1987	3.2	4.6
1988	5.7	4.4
1989	5.4	2.2
1990	3.4	0.7

Source: Economic Trends

SUMMARY OF KEY POINTS

1. The direction of causation in a relationship is largely a matter of theory.
2. An initial step in examining relationships is a careful consideration of the data. Graphs, particularly scatter graphs, are a useful tool in this respect.
3. The line of best fit gives an estimate of the relationship between two variables. This can be drawn freehand, or estimated more carefully using statistical formulae.
4. Often with causal relationships a change in one item affects the other after a period of time has elapsed, i.e. there is a lagged relationship. However, the reasons for such lags and the plausible length of them must be carefully examined.

STUDY F

1. Calculate the correlation coefficient between unemployment and burglary in the UK using the data in Table F1. Put the data on a scatter diagram, and produce a line of best fit. Are there any "outliers"? What effect do they have on the line of best fit?

2. Produce a scatter diagram of money supply growth and inflation levels in different countries. In freehand produce a line of best fit and comment on this. Calculate the regression of inflation against money supply growth and interpret your result. The data is given in Table F2. Redraw the scatter diagram on the assumption that changes in the money supply affect inflation 2 years later. Is this relationship more plausible? Does it work statistically?

3. Using the data in Table F3, produce scatter diagrams of:

(a) clothing expenditure and GDP
(b) clothing expenditure and relative price and comment on possible associations.

Calculate the regression of clothing expenditure against GDP and relative price and interpret your results.

VII QUESTIONNAIRE DESIGN

If, after careful consideration, you have decided to carry out your own survey to collect data, you need to give considerable thought to the design of the questionnaire. It is often said that 'no survey is better than its' questionnaire'. The right questions must be asked to get the required answers. In many circumstances, however, it is quite difficult to ask the 'right' questions. The design of a good questionnaire is an art, not a science. Below are some general prescriptions, but hard and fast rules cannot be given. It is a good idea to get several people to take a critical look at the questionnaire before use. If group work is possible on your the project, this is an area where active participation of all members should be very strongly encouraged.

In this section it is assumed that the questionnaire will be presented at interview, rather than by post. However, many of the points are valid whichever approach is chosen.

1. Questionnaire planning

The scope of the enquiry

A questionnaire should not be used in order to get some responses that 'might be interesting'. Time and thought need to be put in at an early stage to decide what information you need for the project and what information it is practical to obtain from a questionnaire. The right questions need to be asked and the right number of questions. Too long a questionnaire will alienate the respondents; one that is too short will leave you with questions you need to ask unanswered! Always consider how you will use each question in analysing your results – what is its purpose in the overall project?

The type of sample

Stratified quota sampling is commonly used, where the population can be categorised by factors such as sex, age and occupation. A number is interviewed from each category, in order to give a good cross-section of the population. In setting up a survey be careful to use categories that are useful to your research – what groups do you suspect will give different responses? Do not set up too many separate groups, given time and money constraints. (It can take a lot of effort to find people who fit into very narrowly defined categories.)

Pilot survey

Pilot surveys are very useful and should be used wherever the resources permit. The point of the pilot survey is to test the questionnaire and to check that the questions are those that are really needed to pursue the research. The use of a pilot survey does not mean that you can be lax in the initial drawing up of the questionnaire. It does, however, give a chance to correct mistakes that have crept through and to try out some experiments.

Open and pre-coded questions

Pre-coded questions allow answers from a specific list.

Example: *"How do you generally get to work?"*

With answers allowable from the list:

on foot	*1*
by bicycle	*2*
car	*3*
public transport	*4*
other	*5*

This type of question is easy to analyse after the survey, but may, in some situations, not allow a full enough response. Often a mixture of pre-coded and open questions is used. The use of an 'other' category, as in the example above, is usually necessary, both to deal with a range of possible, but minor, replies, not of importance to the inquiry (e.g. by motorbike), and for unforeseen replies (by boat perhaps?). If you think that it may be important to analyse some of these replies, you can add a 'please specify' line.

Ordering

The ordering of the questions can make a difference to the way they are answered. Respondents are known to pay more attention to questions at the beginning and end of a questionnaire and the designer should be aware of this. If respondents are given a list of possible answers, as in pre-coded questions, the alternatives given at the beginning and the end are often favoured, particularly if the list is long. With repeated surveys this problem can be countered by randomising the list.

2. Question design

Using practical questions

(a) information questions

Do not ask questions that the respondent is unlikely to be able to recall.

Example: "How much did your household spend on heating last year?" may be alright to ask in the respondents home, where he/she can check on the information, but is not reasonable in the street.

Do not ask questions the respondent may not know the answer to.

Example: *"How many units of alcohol does each member of your household consume in a week"*, is a poor question if any one person in the household is to give the replies for all.

(b) opinion questions

In asking opinion questions be aware that the respondent may have no opinion, or may have not thought about the question before.

Example: *"Do you agree with the current government's policy towards China?"* Many people may not have considered the policy previously, but may feel driven to give a definite reply by the question being asked.

Views may also be many sided and respondents may have difficulty in fitting their replies into pre-coded answers on a questionnaire. For instance, with the question above, people may agree with the trade policy, but not with the government's reaction to human rights infringements.

Be aware that people may not give accurate answers on personal or emotional matters. This may be deliberate, or simply wishful thinking in response to others perceived opinions. For example, people may underestimate the amount of their alcohol consumption.

(c) classification questions

Classification questions place respondents into different groups by age, sex etc. This type of question is needed in order to stratify the sample and analyse the responses. However, people may object to personal questions on age, occupation etc. It should be made clear to the respondent that the questions are being asked, not out of interest in their personal affairs, but because of the need to classify. Such questions are best put together in a block, at the end or the beginning of the questionnaire.

Avoiding leading questions

(a) with opinion questions

With opinion questions this is a particular problem. The way the question is put may well influence the replies.

Example: The two questions:

"Do you agree that the government was wrong to increase prescription charges"
"Do you agree the government was right to increase prescription charges"

are likely to give a different percentage of respondents saying yes. The standard advice is to be as neutral as possible in the way the questions are phrased.

(b) with information questions

The possibility of leading the respondent should be recognised with factual questions as well, particularly where personal habits are involved. A persons response may be influenced by the assumption in a question that such behaviour is normal.

Example: *"How many books do you read in a month?"* may give the impression to respondents that zero is not seen as acceptable! This can be avoided by use of screening questions where, if the answer is no, the dependent questions are not asked. Here the initial question could be *"Do you regularly read books?"*, although the replies could still be biased in an upward direction if people regard reading as 'good'.

3. Simplicity

The questionnaire should be designed so that it is easy to follow and understand. This involves:

The use of simple language

Always use the simplest word possible. Avoid jargon.

Example: instead of 'terminate' use 'end', use 'tell' rather than 'acquaint'.

Avoid ambiguity and vagueness

There is always a need to check that the question is asking what was intended and that the answers are not some mixture of interpretation.

Example: *"Is your choice of holiday affected by your having children?"*. If this question is asked without first distinguishing whether the respondent has children, or taken recent holidays, the answer 'no' can mean several things. It could mean 'No, I have no children.', or 'I have children but they have not affected my choice of holidays', or even 'No, I have not taken a holiday'.

Exercise 16

Comment on the following questions proposed as part of a survey of people's holiday habits:

(a) *What kind of holiday did you have last year?*
(b) *Would you like to have a holiday abroad every year?*
(c) *How much money has your household spent on holidays this year?*
(d) *Which kind of holiday do you prefer:*
   (i) *skiing*
   (ii) *sun and sea*
   (iii) *mountains*
   (iv) *activity*
   (v) *other*
(e) *A recent study suggested that self catering holidays were better value. Do you agree?*

Organising the questionnaire

The layout and presentation of the questionnaire should make it as easy as possible for the various parties to deal with it – the respondent, the interviewer and the office (that is you processing the results!). If there is a clash between these interests, the respondent should take priority. If respondents are to fill in the questionnaire by themselves particular care should be taken to design the layout in a clear and appealing manner. If the interviewer is to fill in the responses the form can be more directed towards the needs of processing.

SUMMARY OF KEY POINTS

1. Questionnaire design should only be considered after careful project selection and secondary research to establish clearly the role of the primary data collection.
2. In designing the questionnaire very careful attention must be paid to the way questions are structured and phrased, to make them easy and unambiguous for respondents to answer.
3. Do not ask questions which respondents are unlikely to have the knowledge to answer.

4. Do not ask leading questions.

STUDY G

You are part of a team employed by a medium sized travel agency.. The firm is considering expanding into your area, but is concerned to first establish the patterns of holidays taken by local people. Design a questionnaire to find out about the current holiday patterns. You may like to use Table G1 as the basis for developing your own recording sheet, for the numerical part of the survey.

Data supplement

TABLE A1: READING OF NATIONAL NEWSPAPERS: BY SEX & AGE, 1992

	Adult males	Adult females	All adults	15–24yrs	25–44yrs	45–64yrs	65 +	Reader-ship	Readers per copy
	%	%	%	%	%	%	%	millions	numbers
The Sun	24	19	21	28	23	20	15	9.7	2.7
Daily Mirror	19	15	17	18	16	18	17	7.8	2.8
Daily Mail	10	10	10	8	9	11	11	4.5	2.6
Daily Express	9	8	8	7	6	10	11	3.8	2.5
The Daily Telegraph	6	5	6	4	4	7	7	2.5	2.4
Daily Star	7	4	5	8	6	5	2	2.4	3
Today	4	3	3	4	4	3	2	1.5	2.9
The Guardian	3	2	3	3	4	3	1	1.3	3.1
The Independent	3	2	2	2	3	2	1	1.1	2.9
The Times	3	2	2	2	2	3	1	1	2.7
Financial Times	2	1	1	1	2	1	0	0.6	3.6
Any daily newspaper	65	56	60	59	57	64	62	27.3	-
News of the World	29	26	28	35	31	26	19	12.5	2.7
Sunday Mirror	20	18	19	22	20	20	16	8.8	3.2
The People	14	13	13	13	12	15	14	6.1	2.9
The Mail on Sunday	13	13	13	14	14	14	9	5.8	2.9
Sunday Express	11	11	11	10	8	13	14	4.9	2.8
Sunday Times	9	7	8	9	9	8	4	3.5	3
Sunday Telegraph	4	4	4	3	4	5	5	1.8	3.2
The Observer	4	3	4	4	4	4	2	1.7	3.1
Sunday Sport	5	1	3	7	3	1	0	1.3	4.1
Independent on Sunday	3	2	3	4	4	2	1	1.3	3.2
Any Sunday newspaper	71	67	69	71	68	72	66	31.3	-

Percentages of each group reading particular newspapers.

Source: CSO Social Trends 24, HMSO, 1994

TABLE A2: READING OF THE MOST POPULAR MAGAZINES:BY SEX & AGE, 1992

	Adult Males	Adult Females	All adults	15–24	25–44	45–64	65 +	Readership 1992	Readers per copy 1992
	%	%	%	%	%	%	%	Millions	Numbers
Reader's Digest	14	12	13	9	12	17	14	5.9	3.9
Radio Times	13	12	12	14	12	13	12	5.7	3.6
TV Times	11	11	11	15	10	11	9	5	4.5
Viz	14	5	10	29	11	2	0	4.3	1
TV quick	5	8	6	9	8	5	3	2.9	-
What's on TV	6	7	6	9	7	4	5	2.8	1.9
Bella	3	14	9	10	11	8	7	3.3	-
Woman's Own	3	14	9	10	10	8	7	3.4	4.9
Take a Break	4	13	8	12	10	8	4	3	-
Woman	2	11	6	6	7	6	5	2.6	3.8
Woman's Weekly	2	10	6	3	4	7	11	2.4	3.2
Best	2	10	6	7	7	6	3	2.3	3.9

Percentages of each group reading particular magazines.

Source: CSO Social Trends 24, HMSO, 1994

TABLE B1: FOREIGN HOLIDAYS BY UK RESIDENTS: BY DESTINATION

	1981	1986	1989	1990	1981-1990
	Thousands	Thousands	Thousands	Thousands	% change
Belgium/Luxembourg	199	348	278	365	84
Denmark	97	70	62	42	-57
France	2209	2572	3767	4012	82
West Germany	342	415	599	653	91
Greece	881	1472	1566	1551	76
Irish Republic	471	356	580	612	30
Netherlands	310	373	559	593	91
Italy	763	786	891	733	-4
Portugal	362	882	903	885	145
Spain	2841	5556	5717	4637	63
European Community	8474	12830	14922	14083	66
Other Europe	1579	2303	3114	3226	104
Middle East	87	76	78	85	-2
North Africa	175	239	345	278	59
South Africa	29	15	60	46	61
Other Africa	62	83	154	173	178
Australia and New Zealand	27	54	99	101	274
Commonwealth Carribean & Latin America	142	133	269	288	104
United States	717	450	1221	1248	74
Canada	123	86	139	172	40
All Countries	11664	16556	20777	20164	73

Source: CSO, Social Trends, HMSO, 1992

TABLE B2: UNITED KINGDOM: SHARES OF HOUSEHOLD INCOME

	Bottom fifth	Next fifth	Middle fifth	Next fifth	Top fifth	Total
	%	%	%	%	%	%
Gross Income: 1977	9	13	18	24	37	100
Gross Income: 1979	8	13	18	24	37	100
Gross Income: 1981	8	12	17	23	39	100
Gross Income: 1983	8	12	17	23	39	100
Gross Income: 1985	8	12	17	24	40	100
Gross Income: 1987	8	11	16	23	43	100
Gross Income: 1988	7	11	16	23	43	100
Disposable Income: 1977	10	14	18	23	36	100
Disposable Income: 1979	9	13	18	23	36	100
Disposable Income: 1981	9	13	17	23	38	100
Disposable Income: 1983	10	13	17	23	38	100
Disposable Income: 1985	9	13	17	23	38	100
Disposable Income: 1987	8	12	16	23	41	100
Disposable Income: 1988	8	11	16	23	42	100
Post-Tax Income: 1977	9	14	17	23	37	100
Post-Tax Income: 1979	10	13	18	23	37	100
Post-Tax Income: 1981	9	13	17	22	39	100
Post-Tax Income: 1983	9	13	17	22	39	100
Post-Tax Income: 1985	9	13	17	23	39	100
Post-Tax Income: 1987	8	12	16	22	43	100
Post-Tax Income: 1988	7	11	16	22	44	100

Source: CSO, Social Trends, HMSO, 1992

Gross Income = Total income from all sources
Disposable Income = Income after direct taxes
Post-Tax Income = Income after direct and indirect taxes

TABLE B3: GROSS NATIONAL INCOME PER CAPITA

	Brazil	Cote d'Ivoire	Ghana	India	Japan	Rep. of Korea (S. Korea)	Malaysia	United States of America
	US $s	US $s	US $s	US $s	US $s	US $s	US $s	US $s
1971	1260	1000	460	240	12280	1040	940	14790
1972	1390	990	430	240	13170	1080	970	15340
1973	1570	1040	500	240	13970	1220	1140	16070
1974	1640	1050	520	230	13370	1280	1210	15670
1975	1680	1040	450	250	13480	1320	1130	15350
1976	1810	1200	400	250	13920	1500	1310	15980
1977	1860	1370	440	270	14430	1650	1400	16500
1978	1850	1390	460	270	15160	1810	1460	17040
1979	1910	1320	450	250	15640	1900	1620	17300
1980	2000	1280	460	260	15730	1710	1740	16820
1981	1830	1180	420	270	16120	1770	1730	16990
1982	1770	1130	350	280	16520	1900	1750	16450
1983	1670	1070	340	290	16860	2110	1800	16850
1984	1730	1080	360	290	17470	2280	1950	17700
1985	1830	1040	360	300	18260	2410	1830	17980
1986	1980	1070	370	310	19100	2750	1660	18200
1987	2010	900	360	320	19850	3100	1790	18480
1988	1970	830	370	340	20990	3480	1950	19040
1989	1960	760	360	350	21820	3740	2070	19440
1990	1830	670	360	370	22730	4020	2250	19380
1991	—	610	360	370	23730	4320	2380	19000

Source: World Bank, World Tables, 1993

TABLE B4: RETAIL SALES: FOOD, CLOTHING AND FOOTWEAR, 1990=100

	All retailers	Food retailers: total value of sales	Grocers	Dairy retailers and milk delivery	Butchers	Fishmongers	Green-grocers and fruiter-ers	Bread and flour confectioners (Bakers)	Clothing and footwear	Men's and boys' wear	Womens' wear & general clothing, leather & travel	Footwear
1988 Qtr1	79	75	77	92	95	93	88	87	75	79	74	71
1988 Qtr2	82	79	80	101	90	93	99	90	86	88	85	89
1988 Qtr3	85	79	81	100	95	94	85	94	93	96	91	92
1988 Qtr4	104	87	90	103	108	99	85	92	120	126	121	103
1989 Qtr1	84	81	83	100	96	95	91	89	79	84	78	75
1989 Qtr2	88	86	89	106	95	98	106	94	92	91	90	95
1989 Qtr3	89	86	89	104	99	99	90	97	96	95	94	94
1989 Qtr4	110	95	100	112	114	108	93	99	123	125	128	104
1990 Qtr1	90	88	92	99	99	99	98	97	84	86	83	79
1990 Qtr2	95	95	100	110	95	110	109	100	97	91	96	100
1990 Qtr3	96	95	100	109	96	105	94	102	102	98	101	101
1990 Qtr4	116	104	111	116	110	110	94	102	125	124	130	107
1991 Qtr1	95	96	103	100	97	97	97	100	84	82	85	80
1991 Qtr2	100	103	109	114	95	108	107	103	97	85	97	100
1991 Qtr3	102	103	109	113	93	107	98	105	102	93	101	103
1991 Qtr4	120	108	116	116	107	105	95	104	128	119	135	111
1992 Qtr1	99	103	110	98	94	95	100	103	81	74	83	78
1992 Qtr2	104	107	116	98	91	90	111	104	97	83	98	102
1992 Qtr3	104	105	114	99	91	86	91	109	101	87	103	109
1992 Qtr4	125	120	125	100	101	87	87	107	125	117	132	115
1993 Qtr1	105	105	118	98	90	87	95	108	85	69	90	86
1993 Qtr2	110	120	125	94	88	90	102	110	100	81	103	107
1993 Qtr3	111	118	123	92	88	90	89	112	105	85	106	117
1993 Qtr4	132	127	134	96	95	90	82	109	132	120	138	126

Source: *CSO, Monthly Digest of Statistics, HMSO*

TABLE C1: UK GROSS DOMESTIC PRODUCT AT FACTOR COST (£m.@ 1990 prices)

1985 Qtr1	100297
1985 Qtr2	102248
1985 Qtr3	102250
1985 Qtr4	103049
1986 Qtr1	103811
1986 Qtr2	105526
1986 Qtr3	106615
1986 Qtr4	108262
1987 Qtr1	108559
1987 Qtr2	110101
1987 Qtr3	112021
1987 Qtr4	113136
1988 Qtr1	114908
1988 Qtr2	115522
1988 Qtr3	117213
1988 Qtr4	118103
1989 Qtr1	118668
1989 Qtr2	118898
1989 Qtr3	119130
1989 Qtr4	119530
1990 Qtr1	120154
1990 Qtr2	120581
1990 Qtr3	119514
1990 Qtr4	118637
1991 Qtr1	117477
1991 Qtr2	116879
1991 Qtr3	116676
1991 Qtr4	116688
1992 Qtr1	115946
1992 Qtr2	116032
1992 Qtr3	116520
1992 Qtr4	116906
1993 Qtr1	117546
1993 Qtr2	118016
1993 Qtr3	118976
1993 Qtr4	119765

Source: CSO, Monthly Digest of Statistics, HMSO

TABLE C2: HEART DISEASE BY UK REGION PER 100,000 POPULATION, 1990

	MALES per 100,000	FEMALES per 100,000
SOUTH EAST	320	283
EAST ANGLIA	323	279
SOUTH WEST	340	286
EAST MIDLANDS	361	314
WEST MIDLANDS	383	327
YORKSHIRE AND HUMBERSIDE	398	339
WALES	401	331
NORTH WEST	418	362
NORTH	418	376
SCOTLAND	442	385
NORTHERN IRELAND	452	390

Source: CSO, Regional Trends 27, HMSO, 1993

TABLE C3: PASSENGER TRANSPORT USE

	1981	1986	1989	1990	1991
	Bill.Pass.Km	Bill.Pass.Km	Bill.Pass.Km	Bill.Pass.Km	Bill.Pass.Km
Air	3	4	5	5	5
Rail	34	37	40	40	38
Buses and coaches	42	41	41	41	45
Cars, taxis, and two-wheeled motor vehicles	409	471	563	567	596
Bicycles	5	5	5	5	5

Source: CSO, Social Trends 23, HMSO, 1993

Bill.Pass.Km = Billion Passenger Kilometres

TABLE C4: MEMBERSHIP OF EMPLOYERS' AND PERSONAL PENSION SCHEMES

	1988 % curr. emp.	1989 % curr. emp.	1988 % pers. pen.	1989 % pers. pen.
No. employed by current employer: under 3	33	28	18	29
No. employed by current employer: 3 and under 25	41	40	18	28
No. employed by current employer: 25 and under 100	57	59	14	22
No. employed by current employer: 100 and under 1000	72	71	9	13
No. employed by current employer: 1000 and over	83	84	7	11

Source: CSO, Social Trends 22, HMSO 1992

% curr.emp. = percentage of current employees with employers' or personal pension.
% pers.pen. = percentage with personal pension plan.

TABLE D1: PRIVATE HOUSEBUILDING COMPLETIONS: BY NUMBER OF BEDROOMS

	Private enterprise 1 bedroom	Private enterprise 2 bedrooms	Private enterprise 3 bedrooms	Private enterprise 4 or more bedrooms	Private ALL
	'000 houses	'000 houses	'000 houses	'000 houses	'000 houses
1976	6	32	80	21	138
1981	7	24	52	22	105
1986	19	44	62	31	156
1991	19	41	44	33	136
1992	16	38	43	29	125

Source: CSO, Social Trends, HMSO, 1994

TABLE D2: REAL GROSS DOMESTIC PRODUCT AT MARKET PRICES

P.P.S. 1985	West Germany	Italy	United Kingdom	United States of America	Japan
1980	687977	542046	510986	2952045	949786
1981	689175	545034	504271	3019000	986442
1982	682652	546199	512799	2941096	1014448
1983	693309	551488	531547	3055664	1046419
1984	712604	566302	543038	3274529	1098896
1985	726077	581028	562458	3398130	1267981
1986	742242	598003	584545	3506400	1301302
1987	752799	616755	612555	3629916	1354729
1988	780768	641825	638475	3794618	1438828
1989	806117	661294	653067	3900284	1505267
1990	844177	674208	658437	3935344	1589992
1991	872035	681624	646585	3919608	1663130
1992	885988	687759	643352	4021518	1686414
1993	866496	685695	655576	4130099	1684727
1994	866496	696666	671965	4237481	1706629

Source: Statistical Office of the European Communities, EUROSTAT, 1994

PPS 1984 = GDP at 1985 prices, 1985 parities at Purchasing Power Standard

TABLE D3: CONSUMER EXPENDITURE ON DRINK AND TOBACCO

	NOMINAL EXPENDITURE IN CURRENT £ MILLIONS	REAL EXPENDITURE IN CONSTANT 1990 PRICES
	£ million	£m.@ 90prices
1980	14776	31706
1981	16667	30222
1982	17884	28963
1983	19479	29632
1984	20938	29695
1985	22657	29819
1986	23864	29630
1987	25085	29976
1988	26654	30478
1989	27928	30433
1990	30272	30272
1991	33165	29437
1992	34767	28707
1993	36261	28376

Source: *CSO, Monthly Digest of Statistics, HMSO*

TABLE D4: TOTAL POPULATION OF SELECTED COUNTRIES

	West Germany	Italy	United Kingdom	United States of America	Japan
	Millions	Millions	Millions	Millions	Millions
1980	61.3	56.4	56.3	227.7	116.8
1981	61.7	56.5	56.3	230.1	117.6
1982	61.7	56.5	56.3	232.5	118.4
1983	61.5	56.7	56.3	234.8	119.3
1984	61.3	56.9	56.4	237.0	120.0
1985	61.0	57.1	56.5	239.3	120.8
1986	61.1	57.2	56.7	241.6	121.5
1987	61.1	57.3	56.8	243.8	122.1
1988	61.5	57.4	57.1	246.3	122.6
1989	62.1	57.5	57.2	248.2	123.1
1990	63.2	57.6	57.4	250.9	123.5
1991	64.1	57.8	57.6	253.0	123.9
1992	64.9	57.9	57.8	255.4	124.3
1993	65.5	58.0	58.0	258.2	124.7
1994	66.1	58.1	58.2	260.9	125.0

Source: Statistical Office of the European Communities, EUROSTAT, 1994

TABLE E1: LIFE EXPECTANCY IN AFRICA

LIFE EXPECTANCY at birth in years, 1991

	Years @ Birth		Years @ Birth
Mauritius	70	Benin	51
Botswana	68	Sudan	51
Tunisia	67	Tanzania	51
Algeria	66	Zambia	49
South Africa	63	Mali	48
Morocco	63	Burundi	48
Egypt	61	Ethiopia	48
Zimbabwe	60	Burkina Faso	48
Kenya	59	Senegal	48
Namibia	58	Central African Rep.	47
Lesotho	56	Mozambique	47
Ghana	55	Mauritania	47
Cameroon	55	Chad	47
Togo	54	Uganda	46
Gabon	54	Niger	46
Congo	52	Rwanda	46
Cote d'Ivoire	52	Malawi	45
Nigeria	52	Guinea	44
Madagascar	51	Sierra Leone	42

Source: World Bank, World Development Report, 1993

TABLE E2: LIFE EXPECTANCY IN EUROPE (SELECTED COUNTRIES)

LIFE EXPECTANCY: at birth in years, 1991

	Years @ Birth
Hungary	70
Romania	70
Poland	71
Bulgaria	72
Czechoslovakia	72
Portugal	74
Denmark	75
Ireland	75
Austria	76
Belgium	76
Finland	76
Germany	76
France	77
Greece	77
Italy	77
Netherlands	77
Norway	77
Spain	77
Sweden	78

Source: World Bank, World Development Report, 1994

TABLE E3: UK CONSUMERS' EXPENDITURE ON ENERGY £mn 1985 prices

1980 Qtr1	5446	1987 Qtr1	6522
1980 Qtr2	4059	1987 Qtr2	4764
1980 Qtr3	4001	1987 Qtr3	4512
1980 Qtr4	5233	1987 Qtr4	6062
1981 Qtr1	5333	1988 Qtr1	6513
1981 Qtr2	4254	1988 Qtr2	4928
1981 Qtr3	3921	1988 Qtr3	4735
1981 Qtr4	5300	1988 Qtr4	6303
1982 Qtr1	5340	1989 Qtr1	6132
1982 Qtr2	4266	1989 Qtr2	5157
1982 Qtr3	4071	1989 Qtr3	4644
1982 Qtr4	5230	1989 Qtr4	6372
1983 Qtr1	5360	1990 Qtr1	6200
1983 Qtr2	4435	1990 Qtr2	5124
1983 Qtr3	3971	1990 Qtr3	4788
1983 Qtr4	5166	1990 Qtr4	6310
1984 Qtr1	5658	1991 Qtr1	6650
1984 Qtr2	4387	1991 Qtr2	5362
1984 Qtr3	4012	1991 Qtr3	4705
1984 Qtr4	5246	1991 Qtr4	6492
1985 Qtr1	5844	1992 Qtr1	6488
1985 Qtr2	4574	1992 Qtr2	5074
1985 Qtr3	4277	1992 Qtr3	4784
1985 Qtr4	5496	1992 Qtr4	6623
1986 Qtr1	6400	1993 Qtr1	6503
1986 Qtr2	4878	1993 Qtr2	5087
1986 Qtr3	4422	1993 Qtr3	4804
1986 Qtr4	5676	1993 Qtr4	6726

Source: *CSO, Monthly Digest of Statistics, HMSO*

TABLE F1: UNEMPLOYMENT AND DOMESTIC BURGLARY, 1992

	UNEMPLOYMENT RATE %	BURGLARIES PER 10,000 HOUSEHOLDS
	%	per 10,000
NORTH	11.3	3614
YORKSHIRE AND HUMBERSIDE	9.9	3747
EAST MIDLANDS	8.9	2626
EAST ANGLIA	7.6	1941
SOUTH EAST	9.4	2220
SOUTH WEST	9.1	2278
WEST MIDLANDS	10.7	3031
NORTH WEST	10.5	2928
ENGLAND	9.7	2677
WALES	9.7	2235
SCOTLAND	9.5	2214
NORTHERN IRELAND	14.5	1063

Source: CSO, UK Regional Trends, HMSO, 1994

TABLE F2: MONEY SUPPLY GROWTH AND INFLATION RATES

	West Germany	West Germany	Italy	Italy	United Kingdom	United Kingdom	United States of America	United States of America	Japan	Japan
	Money Growth	Inflation	Money Growth	Inflation	Money Growth	Inflation	Money Growth	Inflation	Money Growth	Inflation
1970	9.1	3.3	15.9	4.4	12	6.5	6.6	5.9	16.9	7.6
1971	13.5	5.4	17.2	4.9	16.2	9.7	13.5	4.2	24.3	6.5
1972	14.4	5.5	19	6	23.2	7	13	3.5	24.7	4.8
1973	10.1	7	23.1	10.8	21.8	9.1	6.9	6.2	16.8	11.7
1974	8.5	7	15.7	18.9	10.8	15.9	5.5	11.1	11.5	23
1975	8.6	5.9	23.7	17.3	11.7	24.1	12.6	8.9	16.5	11.8
1976	8.4	4.3	20.8	16.4	11.3	16.6	13.7	5.8	15.4	9.5
1977	11.2	3.7	21.7	18.7	14.8	15.9	10.6	6.4	13.4	8.1
1978	11	2.7	22.6	12.2	15	8.2	8	7.8	14	4.3
1979	6	4.1	20.8	14.6	14.4	13.4	7.8	11.2	10.8	3.7
1980	6.2	5.4	12.7	21.2	17.1	18	8.9	13.5	9.5	7.8
1981	5	6.3	10	17.9	20.4	11.7	10	10.3	11	4.8
1982	7.1	5.4	18	16.5	12	8.6	8.9	6.2	7.9	2.8
1983	5.3	3.2	12.3	14.7	13.2	4.7	12	3.2	7.3	1.8
1984	4.7	2.4	12.1	10.8	13.5	5	8.6	4.3	7.8	2.3
1985	5	2.2	11.1	9.2	13	6	8.3	3.5	8.7	2
1986	6.6	-0.1	10.7	5.8	16.2	3.4	9.4	1.9	9.2	0.6
1987	5.9	0.2	7.2	4.8	16.2	4.2	3.5	3.7	10.8	0.1
1988	6.9	1.3	7.6	5	17.4	4.9	5.5	4.1	10.2	0.7
1989	5.5	2.8	9.9	6.3	18.8	7.8	5	4.8	12	2.3
1990	4.2	2.7	8.2	6.5	12	9.4	3.5	5.4	7.4	3.1
1991	6.3	3.5	9	6.4	5.8	5.9	3	4.2	2.3	3.3

Source: Statistical Office of the Euroepan Communities, EUROSTAT, 1993

TABLE F3: EXPENDITURE ON CLOTHES IN THE UK

	Real Clothing Expenditure	RELATIVE PRICE OF CLOTHES (CLOTHES PI/ CONSUMER PI)	Total Real Personal Disposable Income
	£m.@85prices	1990=100	£m.@90prices
1980	14064	500.1	285411
1981	13934	467.7	283176
1982	14447	445.8	281722
1983	15441	441.1	289204
1984	16261	433.6	299756
1985	17615	430.0	309821
1986	19169	424.6	323742
1987	20204	414.2	334881
1988	20780	408.2	354929
1989	20662	404.6	372356
1990	20876	399.4	380092
1991	20535	382.5	378189
1992	20629	365.7	387646
1993	21481	355.7	393579

Source: CSO, Monthly Digest of Statistics, HMSO

TABLE G1: HOLIDAY SURVEY TEMPLATE

Respondent	Number of people in family	Number of children	Children under the age of 16	No. of holiday stays of 1 night or more away from home	No. of holiday stays of 5 nights or more away from home	No. of foreign holidays	Total number of nights spent holidaying away from home	No. of holidays in Europe (outside UK) in last 5 years	No. of holidays outside Europe in last 5 years
1									
2									
3									
4									
5									
6									
7									
8									
9									
10									